# Synthesis Lectures on Engineering, Science, and Technology

The focus of this series is general topics, and applications about, and for, engineers and scientists on a wide array of applications, methods and advances. Most titles cover subjects such as professional development, education, and study skills, as well as basic introductory undergraduate material and other topics appropriate for a broader and less technical audience.

Abhilash D. Watpade

# From Sand to Chip

## The Beginners' Guide to Semiconductor Manufacturing

Springer

Abhilash D. Watpade
San Diego, CA, USA

ISSN 2690-0300 ISSN 2690-0327 (electronic)
Synthesis Lectures on Engineering, Science, and Technology
ISBN 978-3-032-18438-2 ISBN 978-3-032-18439-9 (eBook)
https://doi.org/10.1007/978-3-032-18439-9



This Springer imprint is published by the registered company Springer Nature Switzerland AG
The registered company address is: Gewerbestrasse 11, 6330 Cham, Switzerland

# Preface

Every day, you use dozens of computer chips without thinking about them. They are in your phone, your car, your coffee maker, and countless other devices. But have you ever wondered how these tiny marvels are actually made?

I work at ASML, a company that makes the machines used to create computer chips. My job takes me around the world to help engineers at chip factories solve complex technical problems with their machines and improve their processes. Over the years, I have noticed something interesting. Even smart people working in technology often do not understand how chips are made. They know their piece of the puzzle, but not the whole picture.

That is why I wrote this book.

The computer chip industry is booming. The United States and other countries are investing billions of dollars to build new chip factories, and companies are racing to create faster, more powerful processors. But there is a problem. We need more people who understand how this technology works. Yet, when someone tries to learn about chip making, they usually find books that are either too technical or too simple.

This book is different. It takes you on a complete journey from ordinary sand to a finished computer chip, explaining each step in plain English. Instead of complex equations, I use everyday examples and analogies. When I talk about growing silicon crystals, I compare it to making rock candy. When I explain how transistors work, I use simple water flow analogies.

This book is for anyone curious about how our digital world actually works. Maybe you are a student wondering about engineering careers. Maybe you work in tech but want to understand the bigger picture. Maybe you are just fascinated by how things are made. Whatever brought you here, you will find explanations that make sense without a technical background.

The chip industry has given us smartphones, computers, the Internet, and technologies that seemed like science fiction just decades ago. Understanding how chips are made helps you understand the foundation of our modern world. And who knows? Maybe this book will inspire you to join this exciting field.

So, let us begin this journey together. By the end, you will never look at your electronic devices the same way again. You will understand the incredible precision and creativity required to transform sand into the processors that power our lives.

Welcome to the amazing world of chip making.

San Diego, CA, USA

Abhilash D. Watpade

# Acknowledgements

Writing this book would not have been possible without the help and support of many people.

First, I want to thank my colleagues at ASML who shared their knowledge and encouraged this project. Working alongside brilliant engineers every day taught me more than any textbook could.

I am grateful to the many engineers and researchers at Intel, TSMC, Samsung, SK Hynix, and other semiconductor companies that patiently answered my questions and helped me understand the finer details of chip manufacturing. Their willingness to share insights from the fab floor made this book much richer and more accurate.

Special thanks to my mentors and professors who first sparked my interest in engineering and taught me that complex ideas can always be explained simply if you understand them well enough.

Thanks goes to my family and friends who listened to endless stories about semiconductors and encouraged me to keep writing, even when I was not sure that anyone would want to read about transistors and clean rooms.

Finally, thanks to all the semiconductor pioneers whose innovations made our digital world possible. From the inventors of the first transistor to today's engineers pushing the limits of physics, your work continues to amaze and inspire.

Any errors or oversimplifications in this book are entirely my own.

**Competing Interests** The author has no competing interests to declare that are relevant to the content of this manuscript.

# Contents

# About the Author

**Abhilash D. Watpade** is an engineer at ASML, where he works on the world's most advanced semiconductor manufacturing equipment, the EUV lithography machines. His job takes him around the globe to help solve complex technical problems at chip factories and mentor new engineers entering the field.

Abhilash holds a degree in mechanical and aerospace engineering from the Illinois Institute of Technology in Chicago and is currently pursuing a PhD in Leadership Studies.

Through his work at ASML, he collaborates daily with engineers at leading semiconductor companies, including TSMC, Intel, Samsung, SK Hynix, and Rapidus. This hands-on experience in the world's most advanced chip factories gives him unique insight into both the technical challenges and the human side of semiconductor manufacturing.

Having seen firsthand how complex chip-making concepts can intimidate newcomers to the field, Abhilash was inspired to write a book that makes this fascinating technology accessible to everyone. He believes that understanding how chips are made should not require a PhD in engineering, just curiosity and the right guide.

When he is not traveling to fabs around the world, Abhilash enjoys sharing his passion for technology and mentoring the next generation of engineers who will continue pushing the boundaries of what is possible with silicon.

# Introduction: The World in a Grain of Sand 1

Have you ever stopped to consider the magic happening inside your smartphone? As you tap the screen to silence your morning alarm, check the weather, or scroll through messages, billions of microscopic switches are firing in precise sequences, transforming your touch into action. These tiny switches, transistors on semiconductor chips, have become the invisible force powering modern life.

## 1.1 Semiconductors in Everyday Life

Your day begins with semiconductors and ends with them too. The alarm that woke you came from a chip in your phone. The coffee maker brewing your morning cup likely contains a microcontroller chip that regulates temperature and timing. As you drive to work, dozens of semiconductor chips control your car's engine, safety systems, and entertainment display. At work, your computer, printer, and the office building's security system all rely on chips. Even devices that seem mechanical, like washing machines or refrigerators, now contain multiple semiconductor chips that manage their operations.

These chips are quite literally everywhere. In medical devices that monitor patients, in the satellites orbiting Earth, in the credit card readers at coffee shops, and in the fitness tracker counting your steps. They've become so integrated into our lives that we barely notice them until they stop working.

But what exactly is a **semiconductor** chip? At its most basic, it's the "brain" of electronic devices, a tiny piece of silicon packed with circuits that can process information. The term "semiconductor" describes materials like silicon that have a unique property: they can sometimes conduct electricity (like metals) and sometimes block it (like insulators), depending on how we manipulate them. Think of a semiconductor as a controllable faucet for electricity, sometimes letting current flow freely, sometimes restricting it to a

 
A. D. Watpade, *From Sand to Chip*, Synthesis Lectures on Engineering, Science, and Technology, https://doi.org/10.1007/978-3-032-18439-9_1

trickle, and sometimes blocking it completely. This controllable behavior is what makes digital computing possible.

The semiconductor age began with a breakthrough moment on December 23, 1947, at Bell Laboratories in New Jersey. There, three scientists, John Bardeen, Walter Brattain, and William Shockley demonstrated the first working **transistor**, a tiny semiconductor device that could amplify electronic signals. This seemingly modest device, which could be held between your fingers, would eventually replace bulky vacuum tubes and transform electronics forever. This innovation made electronic devices dramatically smaller, more reliable, and considerably more energy-efficient. That first germanium transistor (germanium is another semiconductor material used in early transistors) was crude by today's standards, but it sparked a revolution that continues to this day, enabling ever smaller, faster, and more efficient electronic devices (Fig. 1.1).

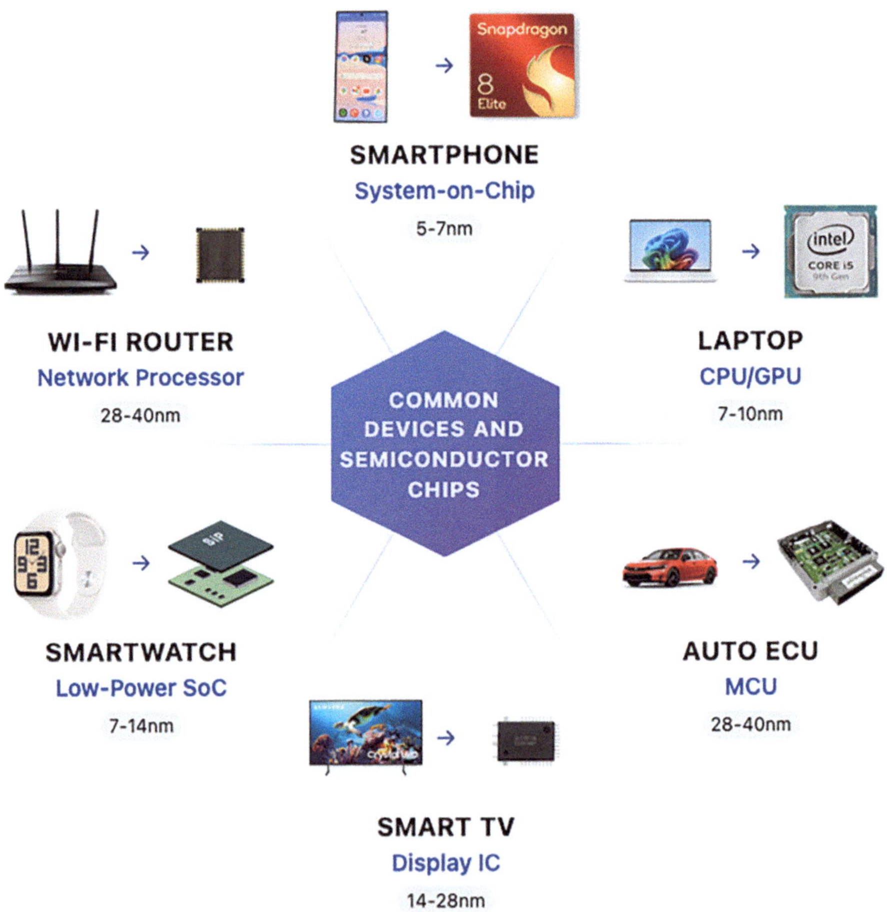


**Fig. 1.1** Common everyday devices and the semiconductor chips that power them

The smartphone in your pocket contains more computing power than all the technology that guided the Apollo missions to the moon, by orders of magnitude. This feat is made possible not by any single breakthrough but by decades of relentless improvement in semiconductor manufacturing. Your typical smartphone includes at least a dozen specialized semiconductor chips: a main processor (CPU), a graphics processor (GPU), memory chips, radio communication chips, power management chips, sensors, and more. Each chip serves a specific purpose, but all are manufactured through variations of the same remarkable process: transforming ordinary sand into extraordinarily complex electronic circuits.

## 1.2 From Sand to Circuit: An Overview of Chip Manufacturing

The journey from sand to sophisticated computing chip is one of humanity's most impressive technical achievements. It begins with ordinary quartz sand (silicon dioxide), one of the most abundant materials on Earth, and ends with some of the most complex objects humans have ever manufactured.

Imagine building a city, but on an almost unimaginably tiny scale. First, you'd need to prepare the land, extracting pure silicon from sand and growing a perfect silicon crystal ingot, which is then sliced into thin, circular wafers. These wafers form the foundation upon which we'll build our microscopic metropolis. Next comes construction: layer by layer, we define and build electronic components like transistors (the buildings), connect them with metal wiring (the roads), and insulate different components from each other (the zoning). Finally, we protect this delicate city under a durable dome (packaging) and connect it to the outside world through specialized interfaces (pins or bumps).

This construction process takes place in specialized facilities called fabrication plants, or "fabs" for short. A modern fab is an engineering marvel in itself, a multi-billion-dollar facility where the air is thousands of times cleaner than a hospital operating room, vibrations are dampened to near-zero, and temperatures are controlled to within a fraction of a degree. The environment must be this pristine because the features being created are so tiny that even a microscopic dust particle can ruin an entire chip.

The process involves hundreds of steps of precisely depositing, patterning, and modifying materials at the atomic scale. Imagine printing billions of perfectly aligned patterns, each one thousands of times thinner than a human hair, stacked in dozens of layers, and doing it consistently, chip after chip, wafer after wafer. This mind-boggling precision happens every day in fabs around the world, producing over a trillion semiconductor chips annually.

The scale is equally astonishing at the individual chip level. A modern processor might contain more than 50 billion transistors, more than six times the human population on Earth, all packed onto a piece of silicon about the size of your thumbnail. And remarkably, every single one of those transistors must work perfectly for the chip to function properly (Fig. 1.2).

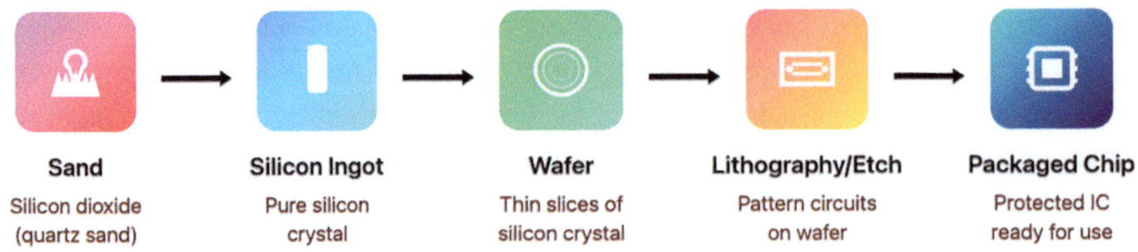


**Fig. 1.2** From sand to chip: Simplified manufacturing flow

## 1.3 The Manufacturing Miracle and Why It Matters

What makes semiconductor manufacturing truly miraculous is not just the end product, but how it has continually redefined what's possible. The transistors on today's chips have features measured in nanometers, billionths of a meter. To put this in perspective, a human hair is about 75,000–100,000 nm thick. Modern chip features can be just 5 nm wide, meaning you could fit about 20,000 transistor gates across the width of a single human hair. At this scale, we are working with components that are just a few dozen atoms wide.

This progression toward smaller, faster, and more efficient chips has followed a remarkably predictable path known as **Moore's Law**. In 1965, Gordon Moore, who would later co-found Intel, observed that the number of transistors on a chip doubled roughly every year while costs decreased. A decade later, the pace settled to about every two years, but the trend held. This wasn't a physical law but rather a goal that the industry set for itself, and astonishingly, it has largely kept pace with this prediction for over half a century. What does this doubling mean in practical terms? Imagine if your bookshelf could hold twice as many books every 2 years while taking up the same amount of space, before long, you'd have an entire library in your living room. Similarly, Moore's Law has enabled devices to become exponentially more powerful while maintaining the same size. Due to this trend, your new phone feels twice as fast as the one from a couple years ago.

The results of this manufacturing revolution are all around us. In 1969, the Apollo Guidance Computer that helped navigate to the moon weighed 70 pounds, consumed 70 watts of power, and could perform about 41,000 instructions per second. Today, the chip in your smartphone weighs just a few grams, is thousands of times more energy-efficient, and can execute billions of instructions per second. This astronomical improvement in performance, size, and energy efficiency, all while reducing cost, is what has enabled the digital revolution.

The impact extends beyond just having faster gadgets. Semiconductor advances have made possible the internet, cloud computing, smartphones, artificial intelligence, modern medical imaging, global positioning systems, and countless other technologies that have transformed how we live, work, and communicate. They've enabled scientific breakthroughs by powering supercomputers that can model climate patterns, discover new drugs, and simulate quantum physics.

The critical importance of semiconductors became starkly apparent during the global chip shortage of 2020–2021. As demand surged for electronics during the COVID-19 pandemic while supply chains were disrupted, manufacturers of everything from cars to game consoles could not get enough chips. Auto factories shut down, product launches were delayed, and consumers faced shortages and higher prices. By 2023, this shortage had eased, but it served as a wake-up call about how essential chips are to the global economy (Fig. 1.3).

Understanding semiconductor manufacturing matters because these chips have become the foundation of modern civilization, as essential to daily life as electricity, clean water, or transportation. Behind every digital experience, from streaming movies to space exploration, lies the remarkable journey from sand to chip.

*The IBM System/360, a 1960s mainframe Computer*

*Apple iPhone 16*

**Fig. 1.3** Computing power evolution: 1960s computer versus modern smartphone (iPhone 16). The IBM System/360, a 1960s mainframe, had processing power up to 16.6 million instructions per second (MIPS) for its high-end Model 91. In contrast, the iPhone 16, released in 2024, uses the A18 semiconductor chip, and can handle billions of instructions per second, making it significantly faster and showcasing the evolution from mainframes to smartphones.Size is also a striking difference. The IBM System/360 was a room-sized mainframe, weighing over 5,000 pounds. The iPhone 16 is a handheld device, fitting in a pocket. This miniaturization is enabled by advancements in semiconductor technology, allowing billions of transistors in a chip smaller than a fingernail

In the chapters that follow, we'll explore each stage of this journey in greater detail, the materials, processes, equipment, and innovations that make it possible. We'll break down complex concepts into understandable parts, revealing the science and engineering that turn simple silicon into the miracle of modern computing. From the growth of pure silicon crystals to the intricate dance of photolithography, from atomic-scale etching to the packaging that protects these delicate circuits, we'll unveil the hidden world inside the devices we depend on every day.

This did not happen overnight. In the next chapter, we'll take a journey through the history of how we got here, from the first transistors to today's nanometer-scale technology.

Throughout this book, we'll see and learn just how that grain of sand is transformed step-by-step into the brain of a computer.

**Key Takeaways**

- Semiconductor chips are the "brains" of modern electronic devices, found everywhere from smartphones and cars to medical equipment and appliances.
- The chip manufacturing process transforms ordinary sand (silicon) into incredibly complex integrated circuits containing billions of transistors through hundreds of precise fabrication steps.
- This manufacturing miracle, evolving along the path of Moore's Law, has enabled the digital revolution, shrinking room-sized computers into powerful pocket devices and creating the foundation for modern technologies that transform how we live, work, and communicate.

**Review Questions**

1. List three electronic devices you used today and describe how semiconductor chips enable their functions.
2. In simple terms, what makes a material a "semiconductor," and why is this property useful for electronic devices?
3. Explain the significance of the first transistor created in 1947 and how it relates to the electronic devices we use today.
4. Moore's Law describes the rapid advancement of semiconductor technology. What specific pattern did Gordon Moore observe, and how has this affected the development of electronic devices over time?
5. Imagine 1 day without semiconductor chips. What daily activities would be impacted and how?

# A Brief History of Semiconductor Manufacturing: From Tubes to Transistors to Today

# 2

The story of semiconductor manufacturing is one of humanity's greatest technological adventures, a journey from clunky vacuum tubes to microscopic transistors that now number in the billions on a single chip. This remarkable transformation happened in just over seven decades, revolutionizing our world more dramatically than perhaps any other technology. It's a tale of brilliant minds, daring companies, and relentless innovation that continues to shape how we live, work, and communicate.

In this chapter, we'll trace this extraordinary timeline, meeting the pioneers who made the breakthrough discoveries and understanding the pivotal moments that transformed electronics from room-sized machines to powerful devices that fit in your pocket. From the first transistor's creation in a New Jersey laboratory to today's cutting-edge chips with features measured in atoms, you'll discover how each innovation built upon the last, creating an industry that now forms the foundation of our digital world.

## 2.1 The Transistor Revolution (1940s–1950s): From Vacuum Tubes to Solid State

In the 1940s, electronics were dominated by vacuum tubes, glass bulbs containing elements that could control electric current flow. These tubes powered everything from radios to early computers like ENIAC (Electronic Numerical Integrator and Computer), which used nearly 18,000 vacuum tubes and consumed enough electricity to dim the lights in Philadelphia when it ran. Though revolutionary for their time, vacuum tubes had serious limitations. They were bulky (roughly the size of light bulbs), generated tremendous heat, consumed massive power, and burned out frequently. A computer might experience a tube failure every few hours, requiring constant maintenance.



A. D. Watpade, *From Sand to Chip*, Synthesis Lectures on Engineering, Science, and Technology, https://doi.org/10.1007/978-3-032-18439-9_2

The world needed something better, and Bell Telephone Laboratories (Bell Labs) was determined to find it. In the winter of 1947, a team of scientists, John Bardeen, Walter Brattain, and William Shockley, was experimenting with semiconductors, materials whose electrical properties fell between conductors (like metals) and insulators (like glass).

On December 23, 1947, Bardeen and Brattain created a makeshift device using a slice of germanium crystal, gold foil, and a plastic triangle. When they applied an electric signal to one part of this contraption, to their amazement, it came out amplified from another part. They rushed to Shockley's office with their strange-looking creation: the world's first transistor.

"That's it! That's it!" Shockley reportedly exclaimed when he understood what they had achieved. This "point-contact" transistor was crude, but it worked, a tiny device that could amplify signals just like vacuum tubes but without the heat, size, or fragility problems. The implications were immediately clear to those who witnessed it: this could transform electronics forever (Fig. 2.1).

News of the transistor spread rapidly throughout the scientific community. Though initially kept quiet for patent reasons, Bell Labs announced the invention to the world in June 1948. The term "transistor", a combination of "transfer" and "resistor", was coined to describe this new device that could switch or amplify electrical signals. For their groundbreaking work, Bardeen, Brattain, and Shockley would later be awarded the Nobel Prize in Physics in 1956.

While the first transistors used germanium, researchers soon realized that silicon, the second most abundant element in Earth's crust, offered better prospects for mass production. In 1954, Gordon Teal at Texas Instruments demonstrated the first commercial silicon

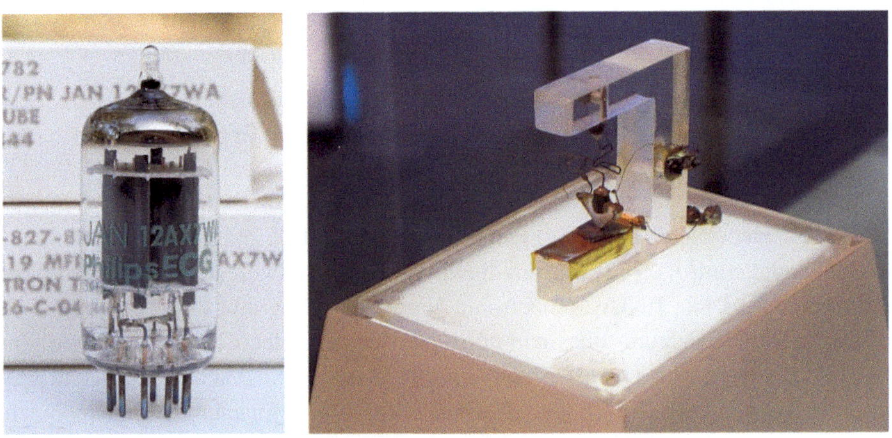

*a. 1940s vacuum tube* *b. The First Transistor ever made, built in 1947*

**Fig. 2.1** The first transistor (point-contact prototype, 1947) compared to a vacuum tube. (**a**) 1940s vacuum tube, (**b**) The First Transistor ever made, built in 1947

transistor. At an electronics conference, he shocked the audience by dunking a working radio with silicon transistors into hot oil, something that would have instantly destroyed germanium transistors. Silicon's superior temperature stability and abundance (essentially refined from ordinary sand) made it the perfect material for the future of electronics.

This period transformed electronics vocabulary, introducing terms like "semiconductor," "transistor," and "solid-state." New companies emerged to capitalize on this technology. William Shockley left Bell Labs to found Shockley Semiconductor Laboratory in 1955, aiming to commercialize transistor technology. Texas Instruments, once an oil exploration company, pivoted to become a semiconductor pioneer. The transistor era had begun, setting the stage for perhaps an even more significant innovation: figuring out how to connect many transistors together on a single chip.

## 2.2 The Integrated Circuit and Planar Process (1960s): Birth of Modern Chip Manufacturing

By the late 1950s, transistors were revolutionizing electronics, but a new problem emerged. Complex electronic devices required dozens or hundreds of transistors, each individually wired together, a tedious, error-prone process called the "tyranny of numbers." Each connection point represented a potential failure, limiting reliability and increasing costs. The next breakthrough would need to solve this connection challenge.

In the summer of 1958, Jack Kilby at Texas Instruments was pondering this problem while most of his colleagues were on vacation. His insight was radical: what if an entire circuit, transistors, resistors, capacitors, and their connections, could be made from the same material and on the same piece of semiconductor? Working alone, Kilby built the first "integrated circuit" or "IC", a simple oscillator circuit on a single germanium chip. When he demonstrated it to company executives in September 1958, showing a single chip generating a continuous signal, they witnessed the birth of modern electronics. Kilby's contribution was so significant that he was awarded the Nobel Prize in Physics in 2000, more than four decades after his invention (Fig. 2.2).

Meanwhile, unknown to Kilby, Robert Noyce at Fairchild Semiconductor was developing a similar idea, but with a crucial manufacturing advantage. In January 1959, Noyce conceived of an integrated circuit using silicon (not germanium) and, critically, incorporating the "planar process" recently invented by his colleague Jean Hoerni.

Hoerni's planar process was ingenious. Previous transistors had components sticking up from the surface with exposed wires, making them fragile. Hoerni's method instead built the transistor in layers on a flat silicon surface, with a protective oxide layer covering everything except where connections were needed. This approach was like building a city from the ground up, layer by layer, rather than constructing each building separately and then trying to connect them.

Noyce realized that with the planar process, he could not only make transistors but also create metal connection paths right on top of the protective oxide layer, connecting

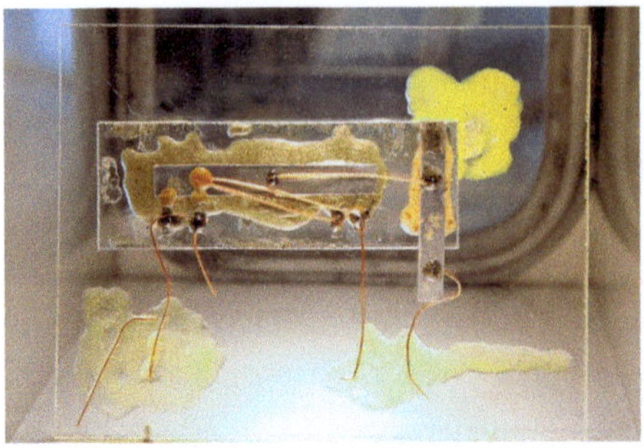

*a. Replica of first integrated circuit by Jack Kilby, 1958*

*b. 4-bit binary counter, Member of the 7400 Transistor-Transistor Logic (TTL) Series; Manufactured by Texas Instruments, 1973*

**Fig. 2.2** Early integrated circuit (1960s): a tiny silicon chip with a few transistors. (**a**) Replica of first integrated circuit by Jack Kilby, 1958, (**b**) 4-bit binary counter, Member of the 7400 Transistor-Transistor Logic (TTL) Series; Manufactured by Texas Instruments, 1973

components without individual wires. This was revolutionary, essentially "printing" circuit connections rather than wiring them by hand.

"Noyce's and Kilby's inventions showed the world that entire electronic circuits could be 'printed' onto a slice of silicon, a profound idea that jump-started the microchip revolution," as industry historian Christophe Lécuyer noted.

The story of the integrated circuit is also the story of Fairchild Semiconductor itself, a company formed when eight talented engineers (later dubbed the "Traitorous Eight" because they left Shockley) departed Shockley Semiconductor in 1957, frustrated by William Shockley's difficult management style. These eight, including Robert Noyce and Jean Hoerni, would transform the valley south of San Francisco into what we now know as Silicon Valley. Noyce would later go on to co-found Intel Corporation with Gordon Moore, further cementing his legacy in semiconductor history.

By the mid-1960s, both Fairchild and Texas Instruments were producing integrated circuits commercially, though they contained just a handful of transistors. An equally important development during this period was the MOSFET (Metal-Oxide-Semiconductor Field-Effect Transistor), invented by Mohamed Atalla and Dawon Kahng at Bell Labs in 1959 and first commercialized in the mid-1960s. Unlike earlier transistor types, MOSFETs were easier to miniaturize and required less power, making them perfect for packing many transistors onto a single chip.

The basic ingredients of modern semiconductor manufacturing were now in place. The planar process for building in layers, photolithography (using light to pattern circuits, covered in Chap. 6) for defining patterns, and MOS transistors for efficient scaling. This was the birth of modern chip manufacturing, where no longer would each transistor be assembled one by one; now they could be made together. It paved the way for complex chips like microprocessors in the 1970s. The stage was set for a period of explosive growth that would transform not just electronics, but society itself.

## 2.3 Moore's Law and the Digital Revolution (1970s–1980s): Scaling Up

In 1965, Fairchild's research director Gordon Moore made an observation that would become the most famous prediction in the history of electronics. In an article for Electronics Magazine, Moore noted that the number of components on integrated circuits had doubled every year since their invention, and he predicted this trend would continue for at least a decade. This forecast, later refined to doubling every 2 years, became known as "Moore's Law."

In simple terms, Moore's Law meant that chips would get twice as complex (twice as many transistors) at roughly the same cost about every 2 years. It was less a physical law than a self-fulfilling prophecy, an industry roadmap that companies strived to follow (Fig. 2.3).

"The complexity for minimum component costs has increased at a rate of roughly a factor of two per year," Moore wrote. "Certainly over the short term this rate can be expected to continue, if not to increase."

The 1970s and 1980s saw this prediction become reality through remarkable engineering achievements. Consider the difference between early and late semiconductor products in this period:

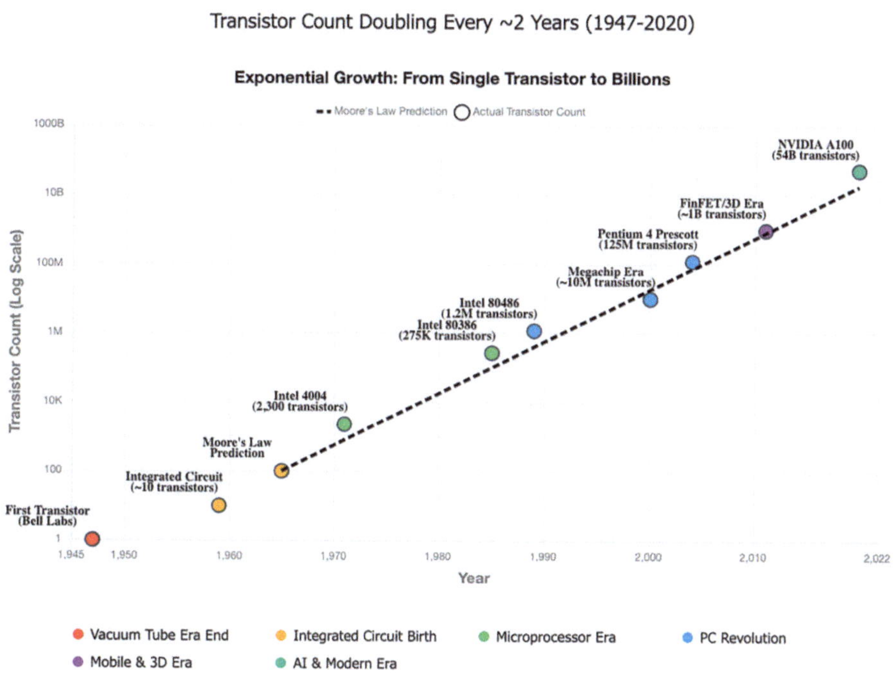


**Fig. 2.3** Trend of transistor counts (Moore's Law) from 1960s to today

In 1971, Intel (a company Moore had co-founded after leaving Fairchild) released the 4004, the world's first microprocessor. This tiny computer-on-a-chip contained 2300 transistors with features about 10 µm wide. By 1989, Intel's 80,486 processor packed 1.2 million transistors with features smaller than 1 µm, a several-hundred-fold increase in complexity in <20 years!

Memory chips followed a similar trajectory. The first commercial DRAM (Dynamic Random Access Memory) in the early 1970s held just 1 kilobit (1024 bits) of data. By the late 1980s, megabit DRAMs storing a million bits were commonplace.

These advances didn't happen by accident. They required continuous innovation in manufacturing technology. Key improvements included:

- Photolithography advanced from using visible light to ultraviolet light with shorter wavelengths, allowing finer pattern printing
- Ion implantation replaced diffusion for more precisely adding dopants to silicon
- Cleanrooms became increasingly particle-free as even microscopic dust could ruin the smaller features
- Computer-controlled equipment improved precision and repeatability

The industry's rapid growth didn't go unchallenged. In the 1980s, Japanese manufacturers like NEC, Toshiba, and Hitachi gained dominance in memory chips, producing them with unprecedented yields and low costs. This competitive pressure forced American companies to innovate or fall behind.

In response, U.S. semiconductor manufacturers formed SEMATECH in 1987, a consortium to advance manufacturing technology. With both industry and government funding, SEMATECH helped American companies regain competitive footing by focusing on manufacturing excellence.

During this period, IBM made critical contributions, including pioneering chemical mechanical planarization (CMP), a polishing technique that would prove essential for building the many layers needed in advanced chips. We'll explore this technology more in Chap. 10.

By the end of the 1980s, the semiconductor industry had transformed modern life. Personal computers were entering homes and businesses, microcontrollers were embedded in appliances and vehicles, and electronic devices were becoming smaller, more powerful, and more affordable every year. The feature sizes on leading-edge chips had shrunk below 1 µm (entering the "sub-micron" era), and chips contained millions of transistors.

Moore's Law had held true, and semiconductor manufacturing had become one of the most advanced engineering disciplines on the planet. Yet even greater transformations lay ahead as the industry globalized and chip manufacturing spread beyond American shores.

## 2.4 The Rise of Global Players (1990s–2000s): From PCs to Mobile and the Foundry Model

The 1990s and 2000s witnessed not just technological evolution but a fundamental reshaping of the semiconductor industry's structure. Perhaps the most significant change was the rise of the "foundry model," which transformed how chips were manufactured globally.

Traditionally, semiconductor companies were integrated device manufacturers (IDMs), they designed and manufactured their own chips, requiring enormous capital for both engineering talent and fabrication facilities. Companies like Intel and IBM exemplified this model, controlling the entire process from design through manufacturing.

In 1987, a new approach emerged when Morris Chang founded Taiwan Semiconductor Manufacturing Company (TSMC). Chang, who had worked at Texas Instruments for 25 years, envisioned a "pure-play foundry", a company that would manufacture chips for other companies but not design its own competing products.

"In the mid-1980s, no one was sure if a company that only makes chips for others could succeed," Chang later recalled. But his vision proved prescient. By 2000, TSMC was manufacturing chips for hundreds of "fabless" companies that focused solely on design, proving the model's worth.

This separation of design and manufacturing democratized the industry. Companies like Qualcomm, NVIDIA, and Broadcom could innovate in chip design without investing billions in manufacturing facilities. Silicon Valley increasingly focused on design, while manufacturing gravitated toward Asia, particularly Taiwan and South Korea, where companies like TSMC, UMC, and Samsung built massive fabrication plants.

On the technology front, several manufacturing breakthroughs defined this era:

In 1997, IBM introduced copper interconnects to replace aluminum wiring in chips. Copper, being a better conductor, enabled faster signals and lower power consumption. As one engineer put it, "To keep improving chips, manufacturers started using copper instead of aluminum for wiring because copper conducts electricity better, meaning faster chips."

Chemical Mechanical Planarization (CMP) became standard practice, allowing manufacturers to polish each layer flat before adding the next, which is essential for building chips with many layers of wiring.

Photolithography continued advancing, with 193 nm deep ultraviolet (DUV) light sources enabling features below 100 nm by the early 2000s. Later, immersion lithography (adding a layer of water between the lens and wafer to improve resolution) was introduced in the mid-2000s, further extending the capabilities of 193 nm lithography systems.

Research on Extreme Ultraviolet (EUV) lithography, using 13.5 nm wavelength light, began in earnest, though it wouldn't reach production until the following decade.

The industry also standardized on larger silicon wafers, 300 mm (12 inches) in diameter instead of the previous 200 mm, in the early 2000s. This allowed more chips to be produced in each manufacturing run, reducing costs per chip.

This period saw explosive growth in chip applications. The 1990s internet boom and subsequent mobile revolution in the 2000s drove unprecedented demand. When Apple introduced the iPhone in 2007, it contained a powerful system-on-chip specified by Apple but manufactured by foundries, exemplifying the new industry model.

The economics of semiconductor manufacturing changed dramatically during this period. A leading-edge fab that cost perhaps $100 million in the 1980s now required billions of dollars, limiting which companies could compete at the cutting edge. This led to consolidation, with fewer companies able to afford leading-edge manufacturing.

By the late 2000s, the industry had transformed into a complex ecosystem of specialized players: fabless design companies, foundry manufacturers, equipment suppliers, and the remaining IDMs. Feature sizes had shrunk to 45 nm and 32 nm, with transistors now measured in tens of nanometers, thousands of times smaller than a human hair. The stage was set for the next phase: true nanoscale manufacturing.

## 2.5 Entering the Nano Era (2010s-Present): Challenges and New Frontiers

As we entered the 2010s, the semiconductor industry faced its most daunting challenges yet. Features had shrunk below 20 nm, where quantum effects and atomic-scale variations become significant obstacles. Moore's Law was getting harder to maintain, requiring increasingly sophisticated innovations to continue scaling.

The most visible innovation was the shift from traditional planar (flat) transistors to three-dimensional designs. In 2011, Intel introduced its 22 nm Tri-Gate technology, later known as FinFET (Fin Field-Effect Transistor). Instead of having the transistor channel lie flat on the silicon surface, FinFETs feature a thin fin of silicon that sticks up from the surface, with the gate wrapped around three sides.

"Think of it like replacing a flat drawing with a 3D model," as one engineer explained. "With the gate controlling the channel from three sides instead of just from above, we get much better control over electrical current flow."

This 3D architecture allowed transistors to continue scaling while maintaining control over leakage current, a crucial improvement as dimensions shrank. FinFETs soon became the industry standard, with TSMC, Samsung, and others adopting similar approaches.

Perhaps the most anticipated technological breakthrough of this period was the arrival of Extreme Ultraviolet (EUV) lithography. After decades of research and development and billions in investment, ASML (the Dutch company that dominates lithography equipment) finally delivered production-ready EUV machines around 2018–2019.

The EUV lithography machines enabled the manufacturing of the 5 nm chips we see today. These massive machines, each costing over $150 million (Newer High NA machines costing over $300 million) and requiring multiple shipping containers to transport, use light with a wavelength of just 13.5 nm to print incredibly small features.

The industry's competitive landscape shifted dramatically during this period. Intel, long the leader in manufacturing technology, stumbled with delays in its 10 nm process. Meanwhile, TSMC and Samsung surged ahead, with TSMC becoming the undisputed manufacturing leader by delivering 7 nm and then 5 nm processes ahead of competitors. This shift highlighted how even the most established companies could struggle with the enormous challenges of advanced manufacturing.

New applications drove demand for increasingly advanced chips. Smartphones grew more powerful with each generation, requiring billions of transistors. Artificial intelligence accelerated the need for specialized processors with enormous computational capabilities. In 2020, NVIDIA's Ampere A100 GPU packed 54 billion transistors on a 7 nm process, a scale that would have seemed almost magical to the inventors of the first transistor.

Engineers jokingly say they're now printing features smaller than the wavelength of light itself, which is true, thanks to clever tricks and multi-patterning techniques!

As we reach the present day, semiconductor manufacturing stands at an unprecedented frontier. Feature sizes on leading-edge chips measure just a few nanometers, dimensions where only a few dozen atoms span a transistor gate. Moore's Law has slowed, with new process nodes taking longer than 2 years to develop, but innovation continues through new materials, 3D stacking, and specialized architectures.

The competition for leadership in making the smallest, fastest chips remains fierce, with TSMC, Samsung, and Intel investing hundreds of billions of dollars to advance their manufacturing capabilities. Each new generation requires more ingenious solutions to continue the story of semiconductor progress, solutions we'll explore in the following chapters of this book (Fig. 2.4).

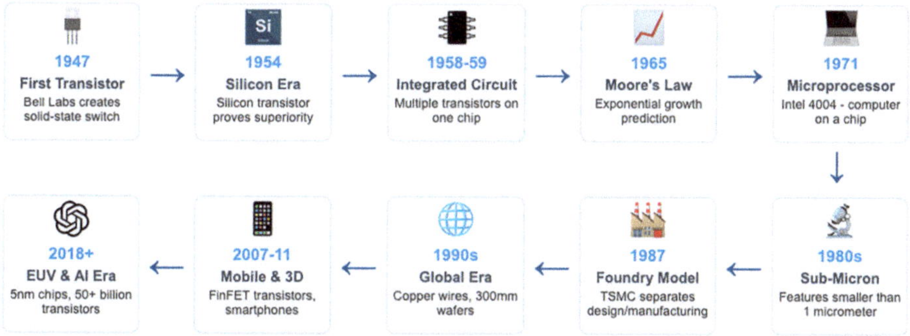


**Fig. 2.4** Key milestones in semiconductor manufacturing from the invention of the transistor to today's advanced chips. Icons from www.flaticon.com (creators: Freepik, flatart_icons)

**Key Takeaways**

- The transistor (1947) replaced vacuum tubes, marking the start of semiconductor electronics. By 1954, the first silicon transistor was made, proving silicon's potential as the foundation material for the industry.
- The integrated circuit (1958–1959) allowed multiple transistors on one chip. The planar manufacturing process and photolithography were key innovations that made modern microchips possible.
- Moore's Law (1965) predicted exponential growth in chip complexity. Throughout the 1970s–1980s, chip manufacturers steadily shrank transistor sizes and improved processes, packing more transistors into chips (from thousands to millions), enabling the digital revolution.
- The 1990s–2000s saw the globalization of chip making: the fabless design and foundry manufacturing model (e.g., TSMC's rise) became prominent. New technologies like copper interconnect, larger 300 mm wafers, and better lithography pushed feature sizes into the sub-micron and nanometer range.
- In the 2010s, chip features entered the low nanometers. 3D transistors (FinFETs) replaced planar ones, and after 20+ years of development, EUV lithography became a reality for cutting-edge chips. Manufacturing leadership shifted among companies, and chips now contain tens of billions of transistors, setting the stage for new challenges and innovations.

**Review Questions**

1. Who were the inventors of the first transistor, and why was the transistor such an important invention for electronics?
2. Explain Moore's Law in your own words. How has this "law" influenced the semiconductor industry?
3. What is the significance of the year 1987 in the context of semiconductor industry business models (hint: think of TSMC)?
4. Name two major changes in chip manufacturing technology that occurred between 1990 and 2020 (e.g., new materials, new transistor designs, or new equipment), and briefly describe why each was important.
5. Which historical development in this chapter did you find most surprising or impactful, and why?

# Semiconductor Fundamentals: Understanding the Material and the Transistor

# 3

Have you ever wondered what makes your smartphone or laptop actually work? Behind every digital device is an extraordinary material with almost magical properties, the semiconductor. In this chapter, we'll explore the fundamental science that makes chips possible, focusing on the special material (silicon) and the remarkable device (the transistor) that together form the building blocks of our digital world.

Unlike other chapters that focus on manufacturing processes, this chapter deals with the "why" behind all those fabrication steps. By understanding these basics, you'll gain insight into what all those carefully orchestrated manufacturing processes are actually trying to create. Think of this as learning about the blueprint before studying how the house is built.

Don't worry if you don't have a background in physics or electronics, we'll keep the explanations straightforward, using everyday analogies instead of complex equations. By the end, you'll understand why silicon is special, how we customize its properties, and how the humble transistor acts as the fundamental switch that enables all computing.

## 3.1 What Makes Silicon Special? Semiconductors vs. Conductors/Insulators

To understand semiconductors, let's start by thinking about materials we're all familiar with. Metal wires conduct electricity easily because they have many free electrons that can move around. When you apply voltage to a copper wire, electrons flow readily; that's electric current. On the other end of the spectrum are insulators like rubber or glass, where electrons are tightly bound to their atoms and can't move freely; that's why rubber gloves protect electricians.


A. D. Watpade, *From Sand to Chip*, Synthesis Lectures on Engineering, Science, and Technology, https://doi.org/10.1007/978-3-032-18439-9_3

Semiconductors like silicon sit in the fascinating middle ground between these extremes. They're neither good conductors nor perfect insulators, but something in between, and more importantly, their behavior can be controlled.

Silicon, the second most abundant element in Earth's crust (after oxygen), forms a crystal where each silicon atom bonds with four neighboring atoms in a precise geometric arrangement. This creates a rigid lattice structure with electrons firmly locked in place.

At very low temperatures, silicon behaves almost like an insulator because all its electrons are tightly bound within these bonds. However, at room temperature, a small number of electrons can gain enough thermal energy to break free from their bonds and move around. Not many, far fewer than in a metal, but enough to allow a small amount of current to flow. This is what makes silicon a "semiconductor"; it conducts, but just a little bit.

Think of silicon as a road with toll gates that are mostly closed. Usually, a few cars (electrons) pass, but if you open the gates or add more lanes, traffic can flow more freely. The remarkable thing about semiconductors is that we can control exactly how well they conduct electricity through various means, especially by adding impurities, a process called "doping" that we'll explore next.

For electrons to move from their bound state to a free state where they can conduct electricity, they need to overcome an energy hurdle called the "bandgap." Think of the bandgap as the minimum energy needed to free an electron from its atom. Silicon's bandgap is just right, not too high (like insulators) and not too low (like metals), making it ideal for electronic control.

Silicon has another remarkable property that proved crucial for electronics. When exposed to oxygen, it forms silicon dioxide ($SiO_2$), basically glass, on its surface. This insulating layer turned out to be extremely useful in making transistors, as we'll see later. It's a bit like having a material that automatically grows its own protective coating exactly where needed.

While silicon wasn't the first semiconductor used (early transistors used germanium), it became dominant because it's abundant (essentially refined from ordinary sand) and because of its convenient oxide. In the 1950s, researchers like Gordon Teal at Texas Instruments championed silicon for these practical reasons, setting the stage for it to become the foundation of the electronics industry.

Other semiconductor materials exist, like germanium, gallium arsenide (GaAs), and silicon carbide, and they're used for specialized applications like high-frequency or high-temperature devices. But silicon remains the workhorse for about 95% of semiconductor devices, thanks to its excellent combination of properties and manufacturability.

## 3.2 Doping: Turning Silicon into an Electrical Conductor on Demand

Pure silicon isn't particularly useful for electronics because it doesn't conduct electricity well enough. The real magic happens when we deliberately add tiny amounts of impurities to silicon, a process called "doping." This simple act transforms silicon's electrical properties in predictable, controllable ways. It's literally what puts the "semi" in semiconductor!

What's remarkable is just how little dopant is needed. Adding just one impurity atom per million silicon atoms can increase conductivity by a factor of a thousand or more! It's like adding a single drop of food coloring to a swimming pool and changing its entire character.

There are two fundamental types of doping, creating two types of semiconductor material:

**N-type silicon:** Add a pinch of phosphorus or arsenic to silicon, and suddenly it has spare electrons that can carry current. These elements have five valence (outer shell) electrons, while silicon has four. When phosphorus replaces a silicon atom in the crystal, four of its electrons form bonds with neighboring silicon atoms, but the fifth is left "extra" and can move freely through the material. We call this n-type silicon (n for negative) because the movable charge carriers are negatively charged electrons.

**P-type silicon:** Alternatively, add a pinch of boron or gallium, and now silicon is hungry for electrons, creating what we call "holes." These elements have only three valence electrons, so when one replaces a silicon atom in the crystal, there's a "missing" electron, a hole, in what should be a complete set of bonds. Nearby electrons can move to fill this hole, creating the effect of the hole moving in the opposite direction. We call this p-type silicon (p for positive) because these holes behave like positive charge carriers.

Think of pure silicon as a semi-full dance floor where not much movement happens. Adding n-type dopant is like introducing a few energetic dancers (electrons) that can move around freely. Adding p-type dopant is like creating empty spots that others rush to fill, creating a chain reaction of movement (holes). Either way, you've suddenly enabled much more activity (Fig. 3.1).

The discovery of these doping effects was crucial to the development of semiconductor electronics. In 1940, Russell Ohl at Bell Labs was experimenting with silicon samples when he noticed that a cracked silicon specimen with different impurities on either side of the crack generated voltage when exposed to light, he had accidentally created the first p-n junction in silicon and discovered the photovoltaic effect in it that makes solar cells possible.

Later, William Shockley and his team at Bell Labs understood that controlling the type and concentration of dopants was essential to making transistors work. This insight turned semiconductors from scientific curiosities into the foundation of modern electronics.

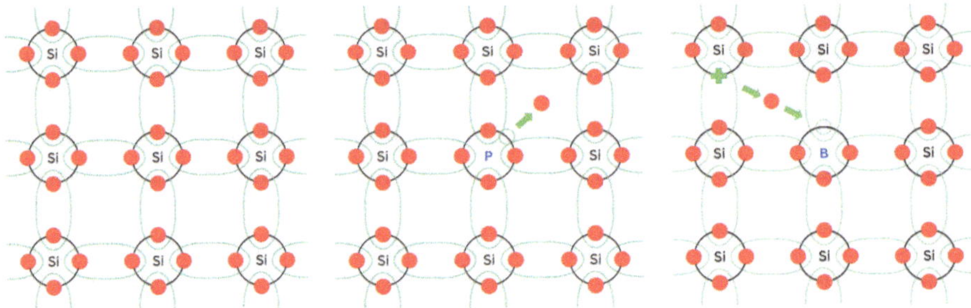


*a] Pure Silicon Crystal*
*A tidy lattice of silicon atoms where every electron is tied up in a bond. Only a handful break free at room temperature, so pure silicon conducts electricity only a little, just enough to call it a semiconductor rather than an insulator.*

*b] n-Type Silicon (Phosphorus-Doped)*
*Swapping in a single phosphorus atom gives the lattice one "spare" electron. That extra electron is free to roam, turning the silicon into an electron-rich (negative, n-type) material that conducts much more easily.*

*c] p-Type Silicon (Boron-Doped)*
*Replacing a silicon atom with boron leaves one bond short an electron, a "hole." Electrons from nearby atoms hop in to fill the gap, so the hole itself moves. This makes the lattice hole-rich (positive, p-type), providing an easy path for current in the opposite direction.*

**Fig. 3.1** Illustration of silicon crystal lattice with n-type (phosphorus) and p-type (boron) dopant atoms. (**a**) Pure Silicon Crystal: A tidy lattice of silicon atoms where every electron is tied up in a bond. Only a handful break free at room temperature, so pure silicon conducts electricity only a little, just enough to call it a semiconductor rather than an insulator. (**b**) n-Type Silicon (Phosphorus-Doped): Swapping in a single phosphorus atom gives the lattice one "spare" electron. That extra electron is free to roam, turning the silicon into an electron-rich (negative, n-type) material that conducts much more easily. (**c**) p-Type Silicon (Boron-Doped): Replacing a silicon atom with boron leaves one bond short an electron, a "hole." Electrons from nearby atoms hop in to fill the gap, so the hole itself moves. This makes the lattice hole-rich (positive, p-type), providing an easy path for current in the opposite direction

In manufacturing, doping can be precisely controlled, we can create regions with specific types and amounts of doping, forming a complex patchwork of differently-behaving silicon on a single chip. This controlled doping is what allows us to create the electronic devices we'll discuss next. Using n- and p-type regions in different arrangements, we can create the basic building blocks of all electronics.

## 3.3 The P-N Junction: The Fundamental Building Block

When p-type and n-type silicon meet, something fascinating happens at their boundary. This p-n junction is perhaps the most fundamental structure in semiconductor electronics, the basis for diodes, transistors, and many other devices.

Let's explore what happens when these two types of doped silicon come together:

Initially, electrons from the n-side (which has extra electrons) rush over to the p-side (which has holes) to fill those holes. This seems like it would create a better conductor where the materials meet, but surprisingly, the opposite occurs.

As electrons and holes recombine at the junction, they neutralize each other and leave behind a zone with no movable charges, called the "depletion region." This region acts like a barrier or wall between the two sides. Without any external voltage, current can't flow freely across this barrier.

Think of it like two different crowds meeting at a fence, one side has extra people, the other side has vacancies. At the fence, they neutralize each other, forming a blockade. Without some external push, movement across this boundary stops.

The truly useful property of a p-n junction emerges when we apply voltage to it:

**Forward bias:** If you connect a battery with its positive terminal to the p-side and negative terminal to the n-side, it pushes holes and electrons toward the junction. This narrows the depletion region, eventually allowing current to flow across. The junction conducts electricity.

**Reverse bias:** If you reverse the battery connections, it pulls holes and electrons away from the junction, widening the depletion region and preventing current flow. The junction blocks electricity.

This one-way behavior, conducting in one direction but not the other, makes the p-n junction act as a one-way valve for electric current. This is precisely what a diode does, the simplest semiconductor device.

Understanding p-n junctions was crucial for William Shockley's team in developing the first transistors. In his groundbreaking 1949 paper, Shockley described how controlling the properties of p-n junctions could create amplifying devices, an insight that led to the junction transistor and eventually the modern electronics era.

The p-n junction's ability to control current flow in one direction forms the basis for many electronic components. Diodes (used in power supplies and signal processing), LEDs (light-emitting diodes), solar cells, and parts of transistors all rely on this fundamental property.

By understanding this basic building block, we're now ready to explore how multiple p-n junctions can be arranged to create the most important electronic component ever invented: the transistor.

## 3.4 The Transistor as a Switch: MOSFET Basics

The modern semiconductor industry is built around a remarkable device: the MOSFET (Metal-Oxide-Semiconductor Field-Effect Transistor). This tiny electronic switch is the fundamental building block of nearly all computing devices today.

At its core, a MOSFET is a tiny switch controlled by voltage. Unlike a mechanical switch that you physically flip, a transistor uses an electric field to turn current on or off, hence the term “field-effect” in its name.

A typical MOSFET has three terminals:

- The **source** (where the current enters)
- The **drain** (where current exits)
- The **gate** (which controls whether current can flow from source to drain)

The basic structure of the most common type (an NMOS transistor) consists of two n-type regions (the source and drain) separated by a p-type region (the substrate). Above this sits a thin insulating layer of silicon dioxide, and on top of that is the metal gate electrode (Fig. 3.2).

Picture two islands of electrons (the n-type source and drain) in a sea of holes (the p-type substrate). Normally, current can’t flow between them because there are two p-n junctions back-to-back, and one of them will always be blocking. But when you apply a positive voltage to the gate, something magical happens: the electric field attracts electrons to the surface of the p-type substrate, creating a thin channel of n-type material connecting the source and drain. With this channel formed, current can flow.

The gate is like a drawbridge controller: when you apply voltage, the drawbridge lowers (forming a channel) so current can flow from source to drain. Remove the voltage, the bridge lifts, and current stops. This on-off behavior is the basis of digital logic, the 1s and 0s that underlie all computing.

There are actually two complementary types of MOSFETs:

- **NMOS** (described above): Has n-type source/drain in a p-type body, and turns on with positive gate voltage
- **PMOS**: Has p-type source/drain in an n-type body, and turns on with negative gate voltage

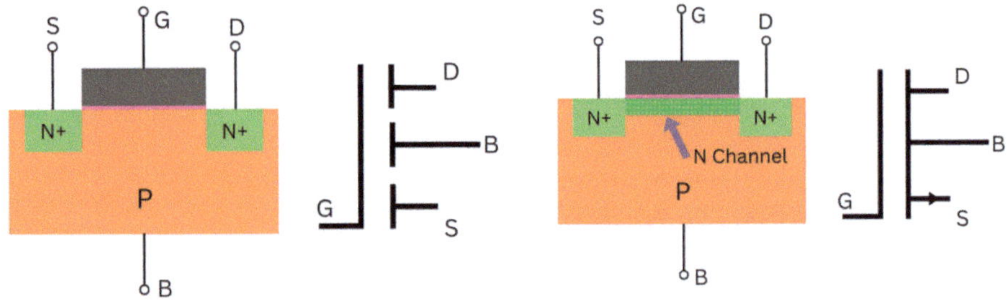


**Fig. 3.2** Cross-section of an NMOS transistor showing source (S), drain (D), gate (G), body (B), oxide layer (pink), and channel formation when the gate is activated

Modern chips use both types together in an arrangement called CMOS (Complementary MOS), which provides efficient operation and low power consumption when idle. By combining NMOS and PMOS transistors, chip designers can create logic gates that form the building blocks of processors, memory, and other integrated circuits.

What makes the MOSFET extraordinary is that the gate is separated from the silicon by an insulator. This means almost no current flows into the gate itself, it controls the flow of current between source and drain using only an electric field, with minimal power needed for the control signal. This property made MOSFETs ideal for integration into complex circuits with millions or billions of transistors (Fig. 3.3).

The MOSFET was invented in 1959 by Mohamed Atalla and Dawon Kahng at Bell Labs, but it took until the late 1960s to become practical for mass production. Today, MOSFETs are manufactured in almost incomprehensible quantities, by some estimates,

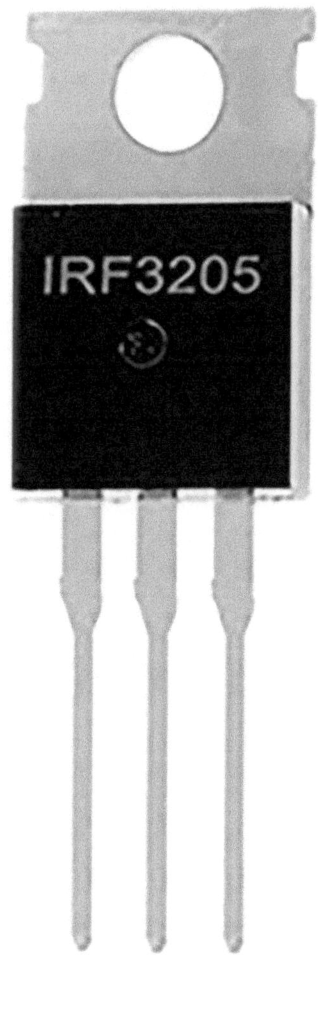


**Fig. 3.3** IRF3205N-Channel Power MOSFET. An actual MOSFET showing source (S), drain (D), and gate (G) pins. Credit: Infineon Technologies AG

over 13 sextillion (13,000,000,000,000,000,000,000) were produced in 2018 alone, making it the most manufactured device in human history.

Modern microprocessors contain billions of these MOSFET switches working in unison, each one manufactured using the precise semiconductor processes we'll explore in the following chapters. Every new generation of manufacturing technology aims to make these transistors smaller, faster, more energy-efficient, and more reliable, pushing the boundaries of what's possible with silicon.

Understanding these transistor fundamentals is crucial because virtually every manufacturing step we'll discuss in later chapters is ultimately about building these tiny switches with precise control and reliable performance. The complexity of semiconductor manufacturing arises from the need to create and connect billions of these transistors with atomic-level precision.

## Summary

- Silicon is the primary material for chips because it's a semiconductor, it can act as an insulator or conductor under different conditions. Its moderate bandgap and ability to form a useful oxide (silicon dioxide) make it ideal for making transistors.
- Doping means adding tiny impurities to pure silicon to change its electrical behavior. Adding phosphorus/arsenic gives extra electrons (n-type silicon), while adding boron creates "holes" (p-type silicon). These charge carriers (electrons and holes) allow current to flow through silicon when doped.
- When p-type and n-type silicon meet, they form a PN junction. A PN junction conducts electricity in one direction but not the other, acting like a diode. This one-way behavior is fundamental in semiconductor devices.
- The MOSFET transistor is essentially a switch built on a semiconductor. It has a gate electrode that, when charged, creates a conductive channel between a source and drain region. In the "on" state, current flows; in the "off" state, it doesn't. This on/off control is the basis of digital logic (0s and 1s).
- Modern chips use billions of MOSFETs (specifically CMOS technology, which pairs NMOS and PMOS transistors) to perform computing tasks. Understanding how a transistor works at a basic level helps in understanding why the manufacturing process has so many steps (to construct these intricate doped regions and layers that form switches).

## Review Questions

1. In simple terms, what is the difference between a conductor, an insulator, and a semiconductor? Give an example of each.
2. What does "doping" silicon mean, and how does it change silicon's ability to conduct electricity?

3. Explain the difference between n-type and p-type semiconductor. What charge carriers does each have?
4. Describe the role of the gate in a MOSFET transistor. What happens when you apply a voltage to the gate of an NMOS transistor?
5. In a MOSFET, why is the insulating layer (oxide) between the gate and silicon so important? What would happen if the gate weren't insulated from the silicon?
6. (Thought exercise) How do you think billions of transistors working together as switches can perform complex computations in a modern processor?

# From Sand to Silicon Wafer: Raw Materials and Crystal Growth

# 4

Before a single transistor can be created or a circuit pattern drawn, we must first produce the perfect foundation, the silicon wafer. In this chapter, we'll trace the remarkable journey of how ordinary sand is transformed into ultra-pure, mirror-polished silicon discs that serve as the starting canvas for chip manufacturing.

This transformation represents one of the most impressive feats of materials engineering in modern technology. We'll go from beach sand to a material so pure that finding an impurity is like locating a single grain of sugar in an Olympic-sized swimming pool, and so flat that any bumps are measured in atoms rather than millimeters.

The processes we'll explore might seem far removed from the intricate circuitry of finished microchips, but they establish the critical foundation upon which all semiconductor devices are built. Without these initial steps, the digital revolution would never have been possible.

## 4.1 Purifying Silicon: From Quartz Sand to Metallurgical and Electronic Grade

Our journey begins with one of Earth's most abundant materials: sand. But not just any sand, we need high-purity quartz sand or quartzite rock, which is essentially silicon dioxide ($SiO_2$). While silicon is the second most abundant element in Earth's crust (after oxygen), in nature it's almost always bound to oxygen, forming silicates or silicon dioxide.

The first challenge is to break those strong silicon-oxygen bonds and extract pure silicon. This starts with a process similar to how iron is extracted from ore in a blast furnace. Quartz sand is mixed with carbon (in the form of coal, coke, or wood chips) and heated to extremely high temperatures, around 1700 °C, in an arc furnace. At this temperature, a chemical reaction occurs:



A. D. Watpade, *From Sand to Chip*, Synthesis Lectures on Engineering, Science, and Technology, https://doi.org/10.1007/978-3-032-18439-9_4

$$SiO_2 + 2C \rightarrow Si + 2CO$$

The oxygen in the sand combines with carbon to form carbon monoxide gas, which bubbles away, leaving behind molten silicon. This silicon is tapped from the furnace and allowed to cool and solidify. What remains is called "metallurgical-grade silicon", about 98–99% pure silicon, with various impurities making up the remainder.

While 98–99% purity might sound impressive, it's nowhere near pure enough for semiconductors. To put this in perspective, imagine you're building a highway where any bump could cause a catastrophic crash. A 99% smooth road would still have potholes every few yards!

For semiconductors to function properly, we need silicon of extraordinary purity, typically "nine nines" (99.9999999% pure) or about one alien atom (impurity) per billion silicon atoms. This level of purity is like ensuring there's only one grain of sugar in a train full of sand, or one incorrect letter in all the books of a major library.

To achieve this extreme purity, metallurgical-grade silicon undergoes a sophisticated purification process. In what's known as the "Siemens process," the silicon is converted into a gas, trichlorosilane ($SiHCl_3$), by reacting it with hydrogen chloride:

$$Si + 3HCl \rightarrow SiHCl_3 + H_2$$

The advantage of converting solid silicon to a gas is that we can then purify it through distillation. Similar to how alcohol is purified in a still, but much more precisely. The trichlorosilane gas is repeatedly distilled to remove impurities that have different boiling points (essentially "cleaning" the silicon by turning it into a gas and back to solid).

Once purified, the gas is then decomposed with hydrogen at high temperature on heated silicon rods:

$$SiHCl_3 + H_2 \rightarrow Si + 3HCl$$

This reaction deposits ultra-pure silicon on the rods, building them up into thick polycrystalline silicon rods of semiconductor-grade purity. The resulting silicon is called "electronic-grade silicon" (EGS) and has impurity levels measured in parts per billion or even parts per trillion.

In the early days of the semiconductor industry, achieving this level of purity was a tremendous challenge. In the 1950s and 1960s, companies like Union Carbide, Monsanto, and Wacker developed these chemical purification techniques, enabling the semiconductor industry to advance beyond its earliest experimental stages.

By the end of this purification process, we have chunks of incredibly pure polycrystalline silicon, silicon that's pure but made up of many small crystals with different orientations. But for semiconductor manufacturing, we need something even more perfect: a single, continuous crystal with atoms arranged in a perfect lattice. That's our next step.

## 4.2 Crystal Growth: The Czochralski Process (Pulling the Silicon Ingot)

Pure silicon is essential, but not sufficient, for semiconductor manufacturing. We also need silicon with a perfect atomic arrangement, which scientists call a "single crystal," where all the atoms align in a continuous, regular pattern throughout the material.

Why is a single crystal so important? Because grain boundaries (where differently oriented crystals meet) would disrupt the flow of electrons and create unpredictable behavior in electronic devices. Imagine trying to drive smoothly across a patchwork of roads built at different angles and heights; that's what electrons would experience in polycrystalline silicon.

To transform our purified polycrystalline silicon into a single crystal, manufacturers use an elegant technique called the Czochralski process, named after Polish scientist Jan Czochralski who discovered it in 1916 (though he was initially working with metals, not silicon).

Picture a large pot of melted silicon, heated to over 1414 °C, hotter than lava. A small, perfect seed crystal of silicon is attached to the end of a rod and carefully lowered until it just touches the surface of the molten silicon. As the seed makes contact with the melt, some of the liquid silicon freezes onto it, adopting the same crystalline pattern as the seed.

Then, very slowly, just millimeters per minute, the rod is pulled upward while rotating. As it rises, more silicon from the melt attaches to the growing crystal, continuing the same atomic pattern. This is very similar to how rock candy is made. You dip a seed crystal (a string or stick) into a saturated sugar solution and slowly pull it out as the sugar crystals grow onto it. The Czochralski process works on the same principle, but with far greater precision and at extreme temperatures (Fig. 4.1).

The process demands exquisite control. Pull too fast, and the crystal may develop defects, rotate at the wrong speed, and the crystal may not grow uniformly. Temperature gradients, rotation rates, and pull speeds are all carefully monitored and adjusted. Even

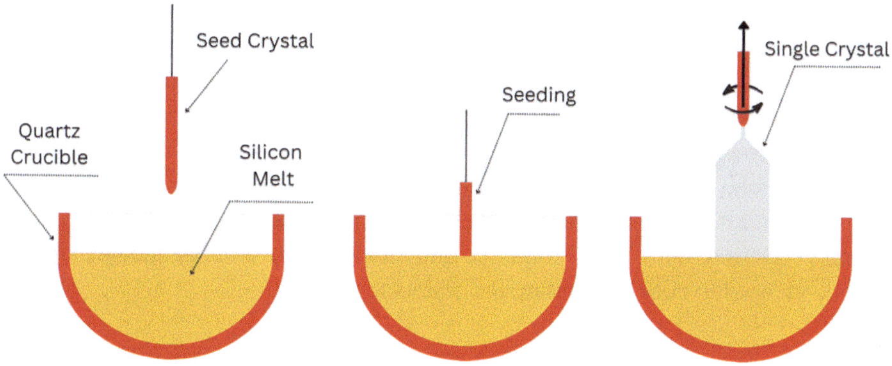


**Fig. 4.1** The Czochralski process for growing a single-crystal silicon ingot, showing the seed crystal being pulled from molten silicon

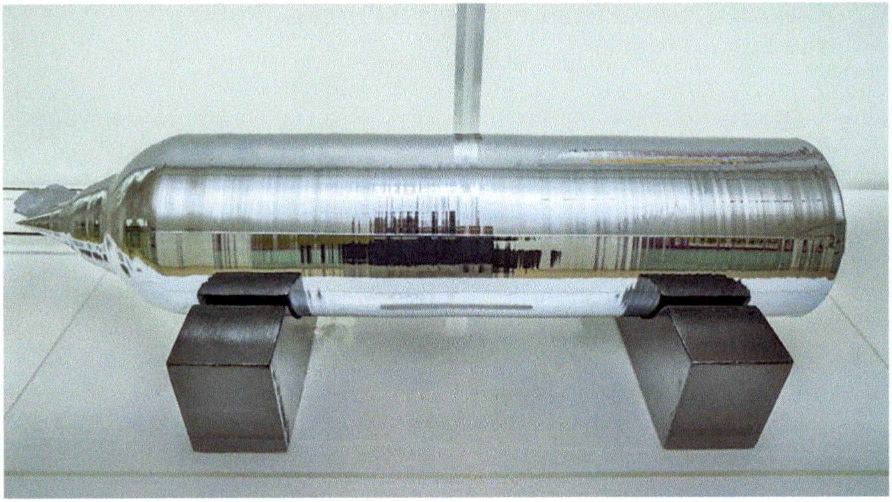

**Fig. 4.2** Actual silicon ingot. Credit: Silicon Materials

vibrations in the building could affect crystal quality, so growth systems are often mounted on massive concrete pads to isolate them.

An interesting technique called "necking" is often used at the beginning of the process. The operators start by pulling a very thin crystal neck, just a few millimeters in diameter. This thin "neck" section is grown to weed out any crystal defects coming from the seed. Once the crystal is perfect, the diameter is increased to form the full ingot.

When complete, a modern silicon ingot is an impressive sight, a shiny, cylindrical crystal up to 300 mm (12 inch) in diameter and 1–2 m long, weighing hundreds of kilograms. These dimensions have grown dramatically over time. In the 1950s, semiconductor manufacturers were working with crystals just a few centimeters wide. By the late 1970s, 100 mm (4 inch) wafers were standard. In the 1990s, 200 mm (8 inch), and since the early 2000s, 300 mm (12 inch) has been the industry standard (Fig. 4.2).

During the crystal growth process, manufacturers can also introduce controlled amounts of dopants, elements like boron (for p-type) or phosphorus (for n-type), into the melt to give the entire crystal specific electrical properties. This background doping provides the foundation upon which more complex doping patterns will later be created.

The entire growth process for a single ingot can take days, with technicians monitoring the process around the clock. It's a delicate art as much as a science, requiring years of experience to master. When complete, we have a silicon ingot of incredible purity and structural perfection, but it's still too large and unwieldy for semiconductor manufacturing. Next, we need to slice it into thin, flat wafers.

## 4.3 Wafer Slicing and Polishing: From Ingot to Mirror-Smooth Disks

With our perfect cylindrical ingot in hand, the next step is to transform it into thin, flat discs, the silicon wafers that will serve as the foundation for chip manufacturing. This process combines precision cutting, grinding, and polishing techniques to create wafers with extraordinary flatness and surface quality.

First, the cylindrical ingot is sliced into thin wafers, much like cutting a salami into slices. Historically, this was done with inner diameter saws (hollow circular blades with diamond cutting edges on the inside), but modern manufacturing typically uses wire saws. In this method, a thin wire coated with diamond slurry is run back and forth through the silicon, cutting multiple wafers simultaneously.

The sawing process is extremely precise, but challenging. Silicon is very hard yet brittle, so cutting must be done carefully to avoid cracks and chips. Modern wafers are remarkably thin, a 300 mm (12 inch) wafer is typically about 775 μm thick, <1 mm. That's roughly the thickness of a credit card, yet with a diameter large enough to cover a dinner plate.

After slicing, the wafers have rough surfaces with microscopic damage from the sawing process. This would be completely unsuitable for the precision work of semiconductor manufacturing, where features are measured in nanometers. Imagine trying to draw an intricate picture on sandpaper instead of smooth paper, the roughness would make precision impossible.

The wafers therefore undergo a series of additional processes to achieve mirror-smooth surfaces:

1. **Lapping**: The wafers are ground with fine abrasives to make them flatter and remove saw marks.
2. **Etching**: Chemical etching with acids like hydrofluoric acid removes a thin layer of silicon damaged during the mechanical processing.
3. **Polishing**: Finally, chemical-mechanical polishing (CMP) creates an ultra-smooth, mirror-like surface. This process combines chemical reactions that slightly soften the surface with mechanical buffing that removes material.

The result is a wafer so flat and smooth that any variations are measured in nanometers (billionths of a meter). To put this in perspective, if we scaled a 300 mm wafer to the size of a football field, the largest bump on its surface would be <1 mm high, smoother than the flattest field you've ever seen (Fig. 4.3).

During the slicing process, the orientation of the crystal lattice is preserved. Each wafer is cut along a specific crystal plane, typically denoted as (100) or (111), because the atomic arrangement affects how transistors will function. To mark this orientation, wafers are given either a small flat edge or a notch. This visual reference helps manufacturers properly align the wafer during subsequent processing steps.

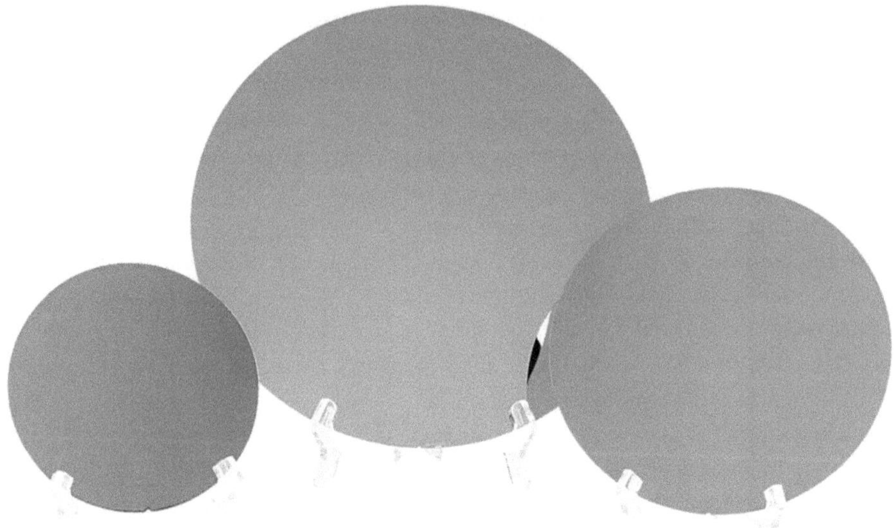

**Fig. 4.3** Polished silicon wafers in different sizes, ready for semiconductor manufacturing, showing their mirror-like surface

After polishing, wafers undergo stringent cleaning processes to remove any residual particles, organic material, or metal contaminants. Even a single microscopic particle can ruin a chip. These pristine wafers are then carefully packaged, often in specialized clean carriers such as cassettes (plastic carriers) or FOUPS, ready to be shipped to chip fabrication facilities ("fabs").

The journey from rough-cut wafer to polished perfection is a testament to precision engineering. The resulting wafers provide the immaculate, atomically smooth surface that serves as the foundation for the intricate nanoscale architecture of modern semiconductor devices.

## 4.4 Wafer Specifications and Evolution: Size, Orientation, and Cleanliness

Before silicon wafers head to chip fabrication facilities, let's take a moment to appreciate just how specialized and carefully specified these shiny discs are. Every aspect of a wafer, from its diameter to its crystal orientation, is precisely controlled to meet the exacting demands of semiconductor manufacturing.

Perhaps the most noticeable change in wafers over the decades has been their increasing diameter:

- 1960s: 1–2 inch (25–50 mm) wafers
- 1970s: 4 inch (100 mm) wafers
- 1980s: 6 inch (150 mm) wafers

- 1990s: 8 inch (200 mm) wafers
- 2000s to present: 12 inch (300 mm) wafers

This growth in size wasn't just for show. Larger wafers allow more chips to be manufactured in a single batch. Since most chip manufacturing steps process the entire wafer at once, larger wafers mean more chips per process run. A 300 mm wafer has more than twice the surface area of a 200 mm wafer, potentially yielding more than double the number of chips for similar processing costs.

The industry has investigated even larger 450 mm (18 inch) wafers since the 2010s, but this transition has been delayed due to the enormous costs of developing new manufacturing equipment and the diminishing returns in cost savings. Many companies continue to use 200 mm wafers for certain types of chips, as the older equipment is fully paid for and economical for some applications.

Crystal orientation is another critical specification. Most wafers are cut so that their surface corresponds to either the (100) or (111) crystal plane. These numbers refer to how the atomic planes in the crystal are oriented. If that sounds technical, just imagine slicing through a perfectly stacked pile of oranges: you could slice horizontally, vertically, or at an angle, and each cut would reveal a different pattern of oranges.

For most logic chips, (100) orientation is preferred because it provides better electrical properties for transistors. Memory chips sometimes use different orientations. The small flat edge or notch on a wafer indicates its orientation and also helps machines automatically align the wafer correctly during processing.

Modern wafers also have precisely controlled thickness and incredible flatness. A 300 mm wafer is typically about 775 μm thick (again, about the thickness of a credit card), with thickness variations of less than a micron. Flatness across the entire wafer is maintained within fractions of a micron, which is essential for the lithography processes that will later create chip patterns.

The edges of wafers are carefully rounded to reduce the risk of chipping and to help with coating processes. These rounded edges also reduce stress concentrations that might cause wafers to crack during thermal processing.

Perhaps most stringent of all are the cleanliness specifications. Before chip processing begins, wafers must be free of virtually all contaminants, metallic impurities, particles, organics, or even improper oxidation. Standard cleaning processes like the "RCA clean" (developed at RCA Laboratories in the 1960s) use sequences of specialized chemical baths to remove different types of contaminants.

The wafer industry is highly specialized, with only a handful of companies worldwide capable of producing semiconductor-grade wafers at scale. Major producers include Japanese companies like Shin-Etsu and SUMCO, GlobalWafers from Taiwan, Siltronic from Germany, and SK Siltron from South Korea. These companies produce tens of millions of wafers annually, supplying the foundation for the global semiconductor industry.

These pristine wafers, pure in composition, perfect in crystal structure, precisely shaped, and immaculately clean, are the starting point for all the fascinating semiconductor manufacturing processes we will explore in subsequent chapters. They represent the literal foundation upon which the entire digital revolution has been built.

Now that we have our pristine wafers, the next step is to take them into the fabrication facility, which we'll explore in Chap. 5.

## Summary

- The base material for chips is ultra-pure silicon, obtained by refining quartz sand ($SiO_2$) through multiple processes. Raw silicon is purified to "nine-nines" purity (99.9999999%) because even tiny impurities can disrupt chip operation.
- Single-crystal silicon ingots are grown using the Czochralski process: a seed crystal is slowly pulled from molten silicon to create a large, cylindrical crystal. This process yields a perfect silicon lattice throughout the ingot, which is crucial for uniform chip behavior.
- The silicon ingot is sliced into thin wafers (<1 mm thick), which are then polished to a mirror finish. The surface needs to be extremely flat and defect-free to support the intricate chip fabrication steps that follow.
- Wafers have standardized sizes, with diameters increasing over the decades from a few inches to the current standard of 300 mm (12 inch). Larger wafers allow more chips per wafer, reducing manufacturing costs per chip.
- Silicon wafers are specified by crystal orientation (e.g., (100) plane) and doping type (p or n background), and they must meet stringent cleanliness standards. These wafers are the pristine "canvas" on which billions of transistors will be patterned in the manufacturing steps to come.

## Review Questions

1. Why is it necessary to have extremely pure silicon for semiconductor manufacturing? What level of purity is typically required for electronic-grade silicon?
2. Briefly describe the Czochralski process of silicon crystal growth. Why do we need a single-crystal silicon ingot rather than a polycrystalline one?
3. How is a silicon ingot turned into wafers? What steps are taken to ensure the wafer surface is flat and clean?
4. List two reasons why larger diameter wafers are beneficial for chip manufacturing. Also, what is the current most common wafer size in industry?
5. What is the purpose of the flat edge or notch on a silicon wafer?

# Inside the Fab: Cleanrooms and the Fabrication Environment

# 5

Before we dive into how chips are actually made, let's step inside the extraordinary environment where this manufacturing magic happens: the semiconductor fabrication facility, or "fab" for short. A fab is unlike any other factory you've ever seen. It's one of the most meticulously controlled environments on Earth, designed to protect microscopic circuits from the invisible threats that surround us every day.

What makes a Fab so special? In this chapter, we'll explore the cleanroom concept that defines semiconductor manufacturing spaces, see how humans suit up to enter these pristine environments, discover how wafers move through a modern fab, and learn about the safety measures that protect both people and products. Understanding this unique environment is crucial before we dive into the manufacturing processes themselves, as it explains why chip manufacturing is so complex, expensive, and fascinating.

## 5.1 The Cleanroom Concept: Defeating Dust and Contamination

Imagine a room so clean that you won't find even a speck of dust in the air. That's what a chip fabrication cleanroom is, far cleaner than a hospital operating theater and far more pristine than the most spotless kitchen you've ever seen. But this extreme cleanliness isn't about aesthetics or comfort, it is an absolute necessity for making working chips.

Why? The features on modern chips are incredibly tiny. Many components measure just a few nanometers, which is thousands of times smaller than a human hair. At this scale even microscopic contamination becomes catastrophic. A single dust particle that you can't even see with your naked eye might be 5–10 μm in diameter. That is large enough to cover thousands of transistors like the ones we learned about in Chap. 3. Think of it like a boulder landing on a miniature highway system. It would completely disrupt traffic.



A. D. Watpade, *From Sand to Chip*, Synthesis Lectures on Engineering, Science, and Technology, https://doi.org/10.1007/978-3-032-18439-9_5

Similarly, a dust particle can short circuit connections or block critical pattern formation or otherwise ruin the chips being manufactured.

This extreme cleanliness directly translates to manufacturing success. Fewer particles mean a higher yield of good chips, the percentage of manufactured chips that actually work correctly.

To combat this invisible enemy, semiconductor fabs maintain extraordinary air quality through several key features:

- **HEPA Filtration**: High-Efficiency Particulate Air filters remove 99.97% of particles 0.3 μm or larger. In a fab, banks of these filters cover the entire ceiling, creating a continuous shower of ultra-clean air.
- **Laminar Airflow**: Unlike turbulent air that swirls and eddies (like in your home), cleanroom air flows in a smooth, "laminar" pattern from ceiling to floor. This sweeps particles downward and away from the production area.
- **Positive Pressure**: Cleanrooms maintain higher air pressure inside than outside. This means that when doors open, air flows outward, preventing outside contaminants from entering. It is similar to what happens when you open an inflated balloon. The air inside rushes out rather than allowing outside air to rush in, just on a room-sized scale and much more controlled.
- **Recirculation**: The air is constantly recycled through filters, with a portion of fresh air added to maintain oxygen levels and remove any buildup of gases or vapors.

Cleanrooms are classified by how clean they are. The traditional U.S. Federal Standard 209E (now replaced by ISO standards) used a simple numbering system: a Class 100 cleanroom allows no more than 100 particles (0.5 μm or larger) per cubic foot of air. For comparison, a typical urban outdoor environment might contain millions of particles per cubic foot! Modern semiconductor processing often requires Class 10 or even Class 1 environments, which is 10 or fewer particles per cubic foot. Some critical process areas, like photolithography, might operate in even cleaner conditions.

To put this in perspective: the cleanest chip fabs are about 10,000 times cleaner than hospital operating rooms. That's like the difference between a dusty attic and a freshly cleaned kitchen!

Early semiconductor manufacturers discovered the critical importance of cleanliness the hard way. In the 1960s, integrated circuit yields were frustratingly low. Many chips simply didn't work. At Fairchild Semiconductor, one of the pioneering chip companies, engineers realized that invisible contamination was killing their devices. By adapting cleanroom techniques originally developed for medical and aerospace applications, they dramatically improved yields. This revelation transformed the industry. No serious chip manufacturer could compete without a proper cleanroom.

Even with all these technological controls, there's one major source of contamination that's particularly difficult to manage: humans. People shed skin flakes, hair, clothing fibers, and exhale particles continuously. A sitting person releases about 100,000 particles per minute, and that number rises dramatically with movement! As Intel and other manufacturers discovered early on, a single human hair or flake of skin could ruin an entire wafer with dozens or hundreds of chips. This leads us to the most visible symbol of semiconductor manufacturing: the cleanroom suit, aka "bunny suit."

## 5.2 Bunny Suits and Protocol: Humans in the Cleanroom

Before entering a semiconductor cleanroom, you must transform. The process begins in a gowning room, where you'll don a special full-body garment known affectionately in the industry as a "bunny suit." This elaborate outfit isn't about protecting you from the cleanroom. It's about protecting the cleanroom from you.

The gowning process follows a careful sequence designed to minimize contamination:

1. First, you step into an "air shower", a small chamber where jets of filtered air blast loose particles from your regular clothes.
2. Next, in the gowning area, you'll cover your shoes with special booties or change into dedicated cleanroom footwear.
3. You will also wear a hairnet and a beard net if necessary.
4. Then comes the main coverall, a one-piece hooded suit made of non-shedding material that covers your entire body except for your face.
5. You'll add a face mask covering your nose and mouth, along with goggles or a full face shield, depending on the cleanroom class.
6. Finally, you put on gloves, making sure they overlap with your suit sleeves, leaving no skin exposed.

The result is a uniform-like garment that makes everyone look nearly identical, with only the eyes visible. The nickname "bunny suit" comes from the hood's shape, which sometimes features rounded protrusions for communication equipment that resemble rabbit ears. After gowning, everyone looks like a high-tech bunny, hence the nickname (Fig. 5.1).

These suits are made from special fabrics designed not to generate particles themselves. The materials are tightly woven and often treated to minimize static electricity, which can attract particles. In the cleanest environments, workers might wear multiple layers or more sophisticated suits.

Even after putting on all this protective gear, cleanroom workers must be mindful of their movements. Quick motions can generate air currents that stir up particles, so people learn to move deliberately and smoothly. Even speaking can release particles, so unnecessary conversation is minimized in the cleanest areas.

Working in a bunny suit isn't particularly comfortable. The suits can be hot, movement is restricted, and breathing through masks all day is tiring. The air in cleanrooms tends to be dry, and the constant airflow can cause dehydration. Many facilities limit how long employees can work in the cleanroom at a stretch due to these challenging conditions.

**Fig. 5.1** Technicians in protective 'bunny suits' in Intel's semiconductor fabrication facility. Credit: Intel Corporation

At major companies like Intel, TSMC, or Samsung, putting on a bunny suit is considered something of a rite of passage for new engineers. The first time can take 15–20 min to properly don all the gear, though experienced workers can suit up in just a few minutes. Visitors to fabs, including executives and dignitaries, must follow the same protocols, no exceptions are made that might risk contamination.

Intel famously featured these distinctive suits in their advertising during the 1990s, with dancing bunny-suited workers in their "Intel Inside" commercials. These ads gave the public a glimpse of the alien-like environment where chips are made, though the actual fab environment is far less conducive to dancing!

Even with all this protective gear, humans are still significant contamination sources, which is why modern fabs increasingly minimize human presence in the cleanest areas. This leads us to how today's fabs are organized and how wafers move through the manufacturing process.

## 5.3 Fab Layout and Wafer Handling: Automation and Flow

A modern semiconductor fabrication facility isn't just one big clean room, it's a carefully organized complex designed to maximize cleanliness while enabling efficient manufacturing. Understanding how a fab is physically arranged helps appreciate the extraordinary lengths the industry goes to for producing perfect chips.

Most fabs use what's called a "bay-and-chase" layout. The manufacturing floor is divided into "bays", the ultra-clean areas where the actual process equipment sits. Behind or beneath these bays are "chases", service corridors where support equipment like pumps, power supplies, and chemical delivery systems are housed. This separation keeps maintenance activities and vibration-generating equipment away from the clean manufacturing space. Only the parts of the tools that directly contact the wafers need to be in the cleanest environment.

In cutting-edge fabs, humans rarely handle wafers directly. Instead, wafers travel through the facility in sealed containers called FOUPs (Front-Opening Unified Pods). A standard FOUP holds 25 wafers in vertical slots, completely protected from the environment. These containers maintain their own mini-environment, only opening when docked to a process tool.

To move these FOUPs between processing stations, modern fabs employ sophisticated automation systems. Overhead transport systems, something like miniature trains running on ceiling-mounted tracks, carry FOUPs from tool to tool. Some fabs use Automated Guided Vehicles (AGVs) that travel along the floor. This automation serves two key purposes. It reduces the risk of human-caused contamination and enables precise tracking of each wafer throughout the complex manufacturing sequence.

The flow of air within the fab is carefully engineered to prevent contamination. In many areas, the floor is perforated with grates to allow air to flow downward, carrying particles away from the process area. Workers often stand on these grates, ensuring that any particles they generate are immediately pulled downward rather than floating toward the wafers.

If you were to tour a fab, one striking feature you might notice is that different areas have different lighting. In particular, photolithography bays, where light-sensitive photoresist is applied and exposed, are illuminated with yellow light only. This special lighting prevents the inadvertent exposure of the photoresist (which is sensitive to blue and ultraviolet wavelengths but not yellow). Walking through these yellow-lit areas feels a bit like entering a photographic darkroom.

The scale of modern leading-edge fabs is staggering. A typical facility might have a clean room area equivalent to several football fields, tens of thousands of square meters. The cost is equally impressive: a new leading-edge fab can cost \$10–20 billion to build and equip. TSMC's latest facilities can exceed \$20 billion, while Intel's recent announcements for new fabs in Arizona and Ohio involve investments approaching \$100 billion across multiple facilities.

These massive fabs operate around the clock, processing thousands of wafers each day. The journey of a single wafer through all the process steps can take 2–3 months and involve hundreds of separate operations. The wafer might travel miles within the fab as it moves from tool to tool. All of this happens in an environment thousands of times cleaner than the outside world, with temperatures controlled to within a fraction of a degree, and vibration dampened to prevent even nanometer-scale disruptions.

## 5.4 Safety and Fab Life: Chemicals, Equipment, and People

While maintaining an ultra-clean environment is critical for making functional chips, semiconductor fabs must also ensure the safety of the people who work there. Chip manufacturing involves numerous hazardous materials and processes that require careful management.

Semiconductor fabrication uses a wide array of dangerous chemicals, strong acids like hydrofluoric acid (which can dissolve glass and bone), toxic gases like arsine and phosphine, flammable substances like silane, and numerous solvents and corrosive materials. These chemicals are essential for the various etching, cleaning, and deposition processes that create chip structures.

To protect workers, fabs have extensive safety systems:

- **Gas detection systems** monitor the air for any trace of toxic gases, automatically triggering alarms and emergency ventilation if detected.
- **Chemical handling areas** have specialized ventilation and often restricted access. Workers who handle these materials directly wear additional protective equipment beyond the standard bunny suit, face shields, chemical-resistant aprons, and specialized gloves.
- **Automated delivery systems** move many chemicals through sealed pipes rather than having humans transport them.
- **Emergency response protocols** are rigorously practiced, including evacuation plans, chemical spill procedures, and first aid training.

Environmental protection is also a major concern. Semiconductor manufacturing uses large volumes of water. A leading-edge fab might use two to four million gallons of ultrapure water daily. This water, along with various chemical wastes, must be treated before returning to the environment. Most fabs have their own water treatment facilities to neutralize chemicals, remove metals, and ensure that wastewater meets or exceeds environmental standards.

The air exhausted from process tools also requires treatment. Scrubbers remove harmful chemicals from the airstream before it's released. Additionally, abatement systems break down greenhouse gases used in some processes (like certain etch steps) to minimize environmental impact.

Despite all the automation, human expertise remains essential in fab operations. Engineers and technicians continuously monitor processes, perform quality checks, and solve problems. Every process has detailed protocols, and workers are trained to follow these procedures exactly to ensure both safety and product quality.

All of these precautions, the pristine environment, careful handling, and precise process control, serve a vital purpose: maximizing yield. In semiconductor manufacturing, "yield" refers to the percentage of chips on a wafer that function correctly. If a wafer contains 500 chips and 450 of them pass all tests, that's a 90% yield.

Every particle, every process variation, every handling error can potentially reduce yield. A single defect might kill just one chip, or it might ruin dozens if it occurs during a critical process step. With modern chips selling for hundreds or thousands of dollars each (for large, complex designs), even small yield improvements translate to millions of dollars in additional revenue.

In the early days of semiconductor manufacturing, yields of 20–30% were considered acceptable. Today, mature processes routinely achieve yields above 90%, with the best exceeding 95%. This improvement reflects decades of refinement in cleanroom technology, automation, and process control, all working together to create an environment where microscopic circuit patterns can be faithfully reproduced billions of times.

Working in a fab requires discipline, attention to detail, and a commitment to following procedures precisely. The reward is participating in one of humanity's most impressive technological achievements, the creation of chips that power our modern digital world.

Now that we've seen how chips are made in an ultraclean, well-controlled environment, we can delve into the processes that turn a blank silicon wafer into an intricate circuit. The first step in that journey is photolithography, a process that uses light to imprint patterns on the wafer, which we'll explore in the next chapter.

## Summary

- Chip manufacturing is done in cleanrooms, highly controlled environments with extremely low dust/particle levels. Even a single dust particle (much larger than chip features) can ruin a device, so the air is continuously filtered and kept particle-free.
- Bunny suits (full-body cleanroom suits) are worn by workers to prevent human contaminants (hair, skin, clothing fibers) from entering the cleanroom. Standard protocols (air showers, gowning procedures) ensure that people carry as little dust as possible into the fab.
- Fabs are heavily automated: Wafers are moved in sealed carriers (FOUPs) by robots and overhead tracks, reducing human handling. Equipment is arranged to maximize cleanliness and efficiency, often with only the critical parts in the clean area and support equipment in service areas. Specialized lighting (yellow lights in lithography areas) is used where necessary to avoid unwanted exposure of light-sensitive materials.
- Maintaining a clean fab is crucial for yield, the percentage of good chips per wafer. A cleaner environment and precise control mean more chips work correctly, which is vital for cost and productivity.
- Fabs also manage safety and hazardous materials carefully. They use toxic gases, strong chemicals, and high temperatures, so there are extensive safety systems (gas detectors, ventilation, protective gear). Fabs have internal facilities to treat waste chemicals and recycle water, ensuring that the high-tech manufacturing is also safe for people and the environment.

## Review Questions

1. What is a cleanroom, and why is it essential in semiconductor fabrication? How does the cleanliness of a chip fab compare to a normal room or a hospital operating room?
2. Why do fab workers wear "bunny suits"? What are these suits designed to protect, and how do they help maintain a clean environment?
3. How do wafers typically move from one machine to another in a modern fab, and why isn't it done by humans carrying them openly?
4. List two safety hazards in a semiconductor fab and briefly explain how fabs mitigate those risks (e.g., handling of toxic gases or corrosive chemicals).
5. (Research/Visit exercise) Find and watch a short video or virtual tour of a semiconductor cleanroom. Write down two new things you learned or observed about the fab environment from the video.

# 6 Photolithography: Printing Patterns on Silicon

Imagine designing the world's most intricate maze, billions of tiny pathways, interconnections, and junctions that must be perfectly arranged for a microchip to function. Now imagine having to transfer that design onto a silicon wafer with precision measured in nanometers (billionths of a meter). How is such an incredible feat accomplished?

The answer is photolithography. It is arguably the most remarkable and critical process in semiconductor manufacturing. Much like a photographer uses light to capture images on film, chip makers use light to print microscopic circuit patterns onto silicon wafers. This chapter explores this fascinating process that essentially turns blank silicon into the foundation of modern computing.

Photolithography serves as the "printing press" of the semiconductor world, enabling manufacturers to create patterns so small that thousands would fit across the width of a human hair. It's a process that combines physics, chemistry, and engineering in a dance of light and shadow that defines our digital age.

## 6.1 Photomasks: The Blueprint of a Chip Pattern

Think of a photomask like a stencil or a photographic negative, it contains the precise pattern we want to transfer onto the silicon wafer. Photomasks are the master templates that define each layer of a chip's intricate circuitry.

At its most basic, a photomask (also called a reticle) is a flat, square piece of high-purity quartz glass with a patterned layer of opaque chrome on one side. The transparent areas of the mask will allow light to pass through, while the chrome-covered areas block light, creating the pattern that will be transferred to the wafer.



A. D. Watpade, *From Sand to Chip*, Synthesis Lectures on Engineering, Science, and Technology, https://doi.org/10.1007/978-3-032-18439-9_6

"If you looked at a mask," explains Dr. Sam, a lithography expert who has worked at Intel, "you'd see a pattern of tiny geometric shapes, which are like the outlines of transistor regions, wires, and other features that will be transferred to silicon."

The pattern on a photomask corresponds to just one layer of the chip, and modern integrated circuits can have 30 or more layers. Thus, a complete set of photomasks (one for each layer) contains the entire blueprint for manufacturing a chip. An advanced processor might require 80 or more masks, with each mask used at a specific step in the manufacturing process (Fig. 6.1).

Modern masks are typically 6 inches square and contain patterns that are 4 times larger than will appear on the final chip. Special reduction lenses in the lithography equipment shrink the pattern by a factor of 4 as they project it onto the wafer. This 4:1 reduction ratio makes mask fabrication somewhat easier, as the features on the mask can be 4 times larger than the final features on the chip.

Creating these photomasks is itself a remarkable process. They're made using specialized tools that use electron beams (rather than light) to write incredibly fine patterns. Since a single defect on a mask could transfer to every chip produced with it, mask quality is absolutely critical. Photomasks undergo intense inspection and cleaning before use.

For the most advanced chips with extremely small features, masks aren't just simple patterns. They often include sophisticated optical correction features that compensate for how light behaves when confined to such tiny dimensions. These corrections are called Optical Proximity Correction (OPC) and phase-shift masks, which help ensure the final pattern on the wafer matches the desired design despite light's tendency to diffract or bend around edges.

Making a set of masks for a cutting-edge chip is a huge endeavor, often costing millions of dollars and taking significant time. But once created, these masks can be used to pattern thousands of wafers. Companies like Toppan, Photronics, and DNP specialize in producing these high-precision photomasks for the semiconductor industry.

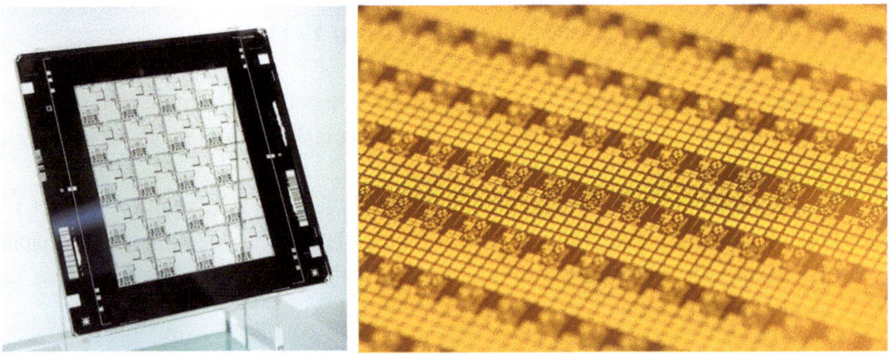

*a) Photomask* *b) Details of structures on chromium photomask.*

**Fig. 6.1** A photomask with chrome pattern (dark) on quartz glass (transparent). (**a**) Photomask. (**b**) Details of structures on chromium photomask

Photomasks have come a long way since the industry's early days. In the 1960s and 1970s, masks were often drawn by hand at a large scale and then photographically reduced. Today, they're created using sophisticated computer-aided design tools and written with electron beam systems that can place features with nanometer precision.

## 6.2 Photoresist Coating: Preparing the Wafer Canvas

Before a pattern can be transferred to the wafer, the silicon surface must first be prepared with a light-sensitive material called photoresist. This remarkable substance is the medium that captures the pattern from the photomask.

Photoresist is essentially a light-sensitive polymer that changes its chemical properties when exposed to certain wavelengths of light. It's a specialized chemical formulation that acts much like the light-sensitive emulsion used in traditional photography.

The process begins with a clean silicon wafer, which is typically already processed with some initial layers. The wafer is placed on a spinner, a high-precision rotating platform. A small amount of liquid photoresist (usually just a few milliliters) is dispensed onto the center of the wafer.

"First, the wafer is placed on a spinner. A few drops of a liquid light-sensitive chemical called photoresist are poured onto the center. Then the wafer rapidly spins, spreading the liquid out into a very thin, even layer across the entire surface, much like spinning a pizza dough or paint splattering evenly when spun," explains Mark Phillips, a photolithography specialist.

The spinner rotates at thousands of revolutions per minute, and the centrifugal force causes the photoresist to spread evenly across the wafer's surface. The final thickness of the photoresist layer is carefully controlled by the spin speed and photoresist viscosity, typically resulting in a film between 100 and 500 nm thick and about 1/200th the thickness of a human hair (Fig. 6.2).

After spinning, the wafer undergoes a "soft bake", a brief heating step that evaporates the solvents in the photoresist and solidifies the film. This creates a stable, even coating across the wafer.

There are two main types of photoresist, each with different behavior when exposed to light:

**Positive photoresist**: When exposed to light, the chemical bonds in the resist break down, making the exposed areas more soluble in the developer solution. The light essentially "weakens" the resist. After development, the areas that were exposed to light are washed away, while the unexposed areas remain intact. Think of positive resist as "what you see is what you remove", expose it, and that part goes away.

**Negative photoresist**: This works in the opposite way. Light exposure causes the resist to polymerize and harden, becoming less soluble. After development, the unexposed areas

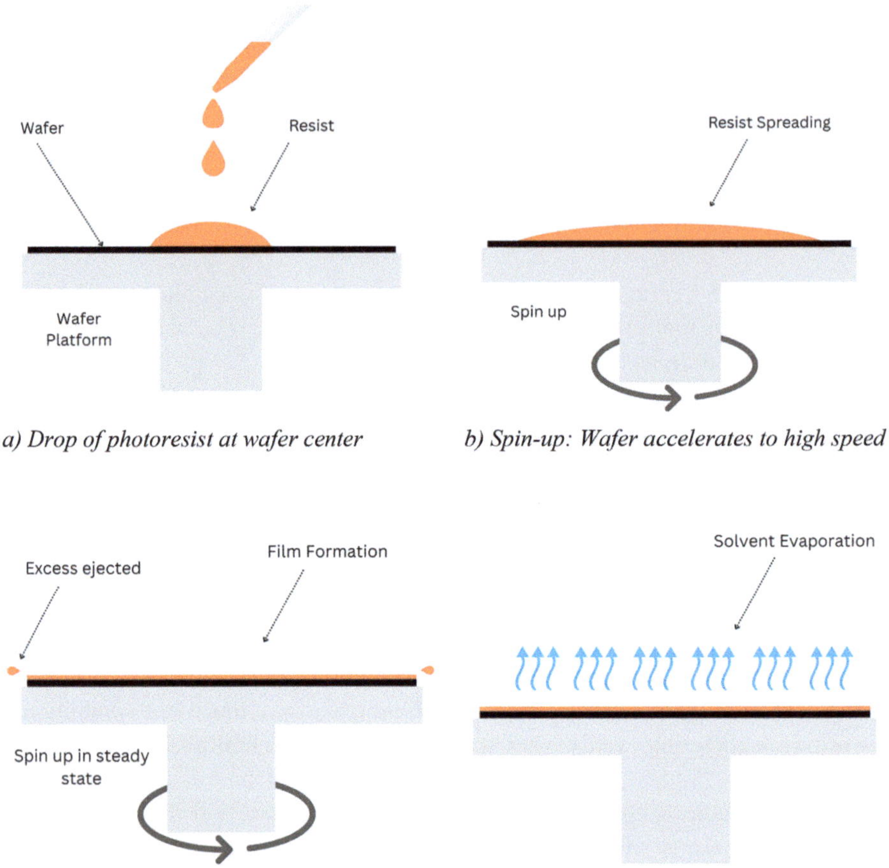


*a) Drop of photoresist at wafer center*

*b) Spin-up: Wafer accelerates to high speed*

*c) Radial Spreading: Photoresist forms a thin, even layer as excess is ejected*

*d) Drying: Solvent evaporates, leaving a uniform photoresist film*

**Fig. 6.2** Wafer spin coating process showing photoresist spreading during high-speed rotation], (**a**) Drop of photoresist at wafer center, (**b**) Spin-up: Wafer accelerates to high speed, (**c**) Radial Spreading: Photoresist forms a thin, even layer as excess is ejected, (**d**) Drying: Solvent evaporates, leaving a uniform photoresist film

wash away, while the light-exposed regions remain. With negative resist, what you expose stays (unexposed areas wash away).

Most modern semiconductor manufacturing uses positive photoresist due to its superior resolution capabilities for very small features. For the remainder of our discussion, we'll focus on the more common positive resist process.

In a positive resist, the areas exposed to light will wash away during development, so light essentially 'prints' the pattern by removing resist. It's similar to old photographic film developing, where light changes the material so a developer solution can reveal the pattern.

The quality and uniformity of this photoresist layer are critically important. Any variation in thickness across the wafer could lead to focus problems during the exposure step, resulting in pattern distortion. The coating must be free of bubbles, particles, and other defects that could transfer to the final pattern.

At this point, the wafer is uniformly coated with a light-sensitive film, ready for the next crucial step, exposure to light through the photomask to create the desired pattern.

## 6.3 Exposure and Development: Transferring the Pattern

With the wafer coated in photoresist, we reach the heart of the photolithography process: exposure and development. This is where the magic happens, and the pattern from the photomask is transferred to the wafer.

### 6.3.1 Exposure

The resist-coated wafer is loaded into a sophisticated piece of equipment called a stepper or scanner (formally called a “step-and-repeat” or “step-and-scan” system). These machines are among the most complex and expensive in the semiconductor fab, with advanced models costing well over $100 million each.

The machine positions the wafer under a mask that contains the desired pattern. A strong UV light shines through the mask. Wherever the mask has clear openings, the light passes through and reaches the photoresist, altering it. In areas where the mask is dark, no light hits the resist.

The light source used has evolved over the decades. Early systems used mercury arc lamps that produced ultraviolet light with wavelengths around 365 nm (called i-line lithography). As chips got smaller, manufacturers moved to excimer lasers producing deeper ultraviolet light, first at 248 nm (using krypton fluoride, or KrF) and then at 193 nm (using argon fluoride, or ArF) (Fig. 6.3).

The pattern from the photomask is projected through a sophisticated optical system that focuses and (typically) reduces the image onto the wafer. Modern systems can project extremely sharp images with features measured in nanometers across the surface of the wafer.

A key challenge is alignment between layers. Each new layer must align precisely with the layers already on the wafer. Advanced lithography systems use laser-based alignment systems that can detect special alignment marks on previous layers and position the wafer with nanometer precision. Even a 10 nm misalignment could mean a via or contact misses the transistor it should connect to, potentially causing the chip to fail. This is why

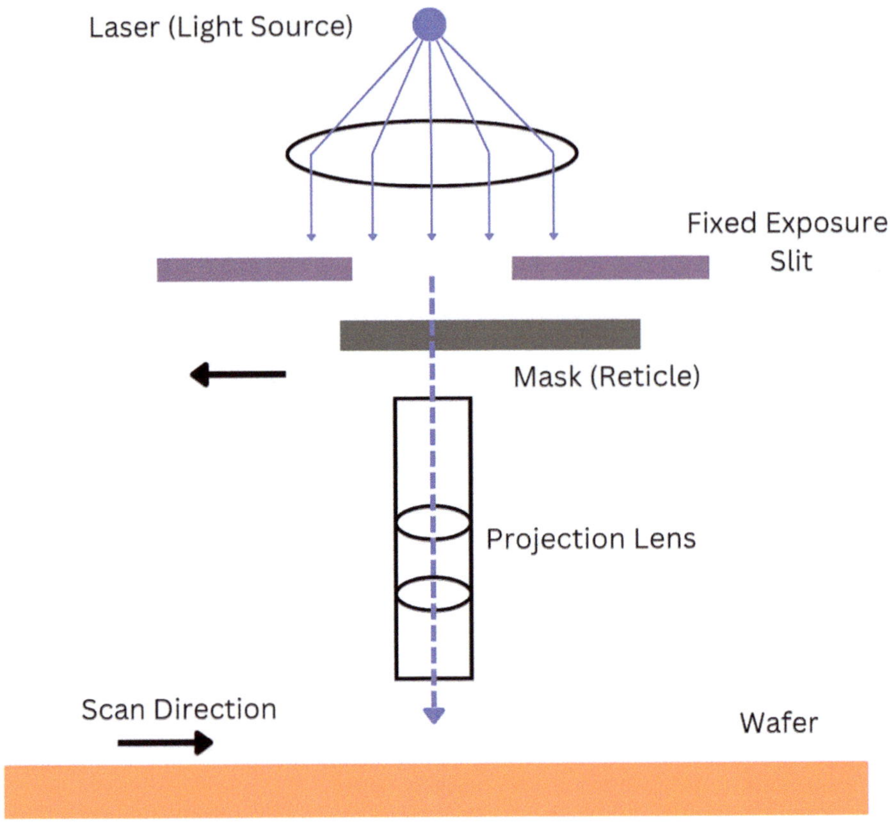


**Fig. 6.3** Schematic diagram of the exposure process showing light source, mask, lens system, and wafer

alignment systems use lasers and marks on the wafer to position each layer within a few nanometers of where it should be.

"The machine also carefully aligns each pattern to the previous patterns on that wafer, so that all layers of the chip line up perfectly," notes Dr. Tsu from a major equipment manufacturer. "Modern tools achieve alignment accuracy of just a few nanometers across a 300 mm wafer, and it's like hitting a target the size of a quarter from a mile away."

Since the pattern for a single chip is much smaller than the wafer, the exposure process is repeated across the wafer, moving (or "stepping") from position to position until the entire wafer is exposed. This is why the machines are called "steppers" as they step across the wafer, exposing one chip area at a time.

### 6.3.2 Development

After exposure, the wafer is moved to a development system. Here, a developer solution (typically a mild alkaline solution) is applied to the wafer.

"After exposure, the wafer is put into a developer solution (similar to how camera film is developed). The developer dissolves the parts of the resist that were exposed to light (for a positive resist)," explains Dr. Willson, a Process Engineer at a leading Fab who has worked on photoresist formulations.

As the developer solution works, it dissolves away the areas of positive resist that were exposed to light, leaving behind the unexposed areas. The result is a pattern of photoresist on the wafer that corresponds to the opaque areas on the photomask. The developer is then rinsed away with deionized water, and the wafer is dried.

The patterned photoresist now serves as a protective mask for subsequent processing steps. Wherever the resist remains, the underlying materials will be protected. Where the resist has been removed, the materials will be exposed to the next process (such as etching, implantation, or deposition).

The resist that remains acts as a protective stencil for the next steps, covering parts of the wafer we don't want to change and leaving other parts open for processing," says Dr. Willson.

### 6.3.3 Resolution Limits

A critical parameter in photolithography is resolution, i.e., the smallest feature size that can be reliably printed. This is fundamentally limited by the wavelength of light used for exposure. As a rule of thumb, the smallest feature that can be printed is roughly half the wavelength of the light used. However, this also depends on other factors like the numerical aperture of the lens system and process optimizations.

The push for smaller and smaller transistors has driven the industry to use shorter wavelengths of light and to develop ingenious techniques that extend the resolution limits of each lithography technology. This continuous improvement in resolution is what has enabled Moore's Law, the doubling of transistor density every 2 years, to continue for decades.

## 6.4 Pushing the Limits: Immersion Lithography and EUV

As transistors shrank below 100 nm, the semiconductor industry faced a serious challenge. The wavelength of light used for lithography was becoming a fundamental limitation to further miniaturization. This section explores two revolutionary technologies that helped overcome these limits: immersion lithography and extreme ultraviolet (EUV) lithography.

### 6.4.1 Immersion Lithography: Water to the Rescue

By the mid-2000s, chipmakers were pushing 193 nm lithography to its limits. The industry needed to pattern features smaller than the wavelength of light being used, which was a seemingly impossible task.

"When chip features became smaller than the wavelength of light we used, engineers got creative. One trick was immersion lithography: they put a drop of ultra-pure water between the lens and the wafer," explains Dr. Nakagawa, who worked on early immersion systems. "Because light travels differently in water than air (slower, with a shorter effective wavelength), this allowed the patterns to be focused more tightly, printing smaller features than dry light could. It's like using water as a magnifying lens."

The concept sounds deceptively simple. Replace the air gap between the final lens element and the wafer with purified water. Water has a higher refractive index than air (1.44 vs 1.0), which effectively reduces the wavelength of the 193 nm light to about 134 nm in the water. This allowed for tighter focusing and smaller feature sizes, about a 30% improvement in resolution (Fig. 6.4).

But implementing immersion lithography was far from simple. Engineers had to solve numerous challenges:

- Keeping the water ultra-pure (even tiny contaminants could cause defects)
- Preventing bubbles from forming in the water (which would distort the light)
- Ensuring the water stayed in place as the wafer moved rapidly
- Dealing with water-related defects like "watermarks"

Despite these challenges, immersion lithography was successfully implemented around 2007 and became the industry standard for advanced nodes. This technology extended the life of 193 nm lithography far beyond what many thought possible, enabling the 45 nm,

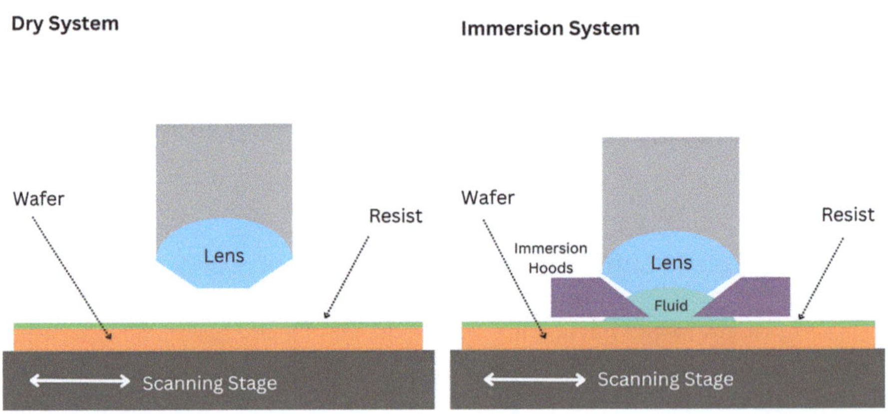


**Fig. 6.4** Comparison between conventional dry lithography and immersion lithography with water between the lens and the wafer

32 nm, 22 nm, and even 14 nm technology nodes with the help of additional techniques like multiple patterning (exposing the wafer multiple times to create more complex patterns).

Immersion lithography gave 193 nm light another decade of life in leading-edge manufacturing. It was a crucial bridge technology while the industry worked on the next big thing: EUV.

### 6.4.2 EUV Lithography: A 20-Year Moonshot

Even with immersion and multiple patterning, there were limits to how far 193 nm lithography could be stretched. The industry needed a fundamentally new approach, and found it in Extreme Ultraviolet (EUV) lithography.

Eventually, even tricks like immersion weren't enough, so the industry turned to a fundamentally shorter wavelength: Extreme Ultraviolet (EUV) light, at 13.5 nm and about 1/14th the wavelength of the previous light. EUV can create even tinier patterns, but it comes with huge challenges. This light is absorbed by nearly everything (including air), so the whole process must happen in a vacuum with special mirrors instead of lenses.

The move to EUV lithography represented a radical change:

- The wavelength jumped from 193 nm to just 13.5 nm
- Traditional lenses couldn't be used (since EUV is absorbed by glass), so special reflective optics with multilayer mirrors were required
- The entire light path had to be in a vacuum
- Conventional transmission masks wouldn't work, requiring the development of reflective masks
- Generating enough EUV power was extraordinarily difficult

The story of EUV development is one of the most remarkable engineering journeys in semiconductor history. The Dutch company ASML led this effort, working with partners like Intel, Samsung, TSMC, and research institutions worldwide. The development spanned over 20 years and cost billions of dollars.

For two decades, scientists and engineers from ASML and its partners worked to make EUV practical. They had to invent new light sources, develop special mirrors that could reflect this elusive wavelength, and solve countless other problems. Many doubted it would ever succeed.

One of the biggest challenges was creating a powerful enough source of EUV light. The solution was ingenious yet extraordinarily complex. Tiny droplets of tin are shot into a vacuum chamber, where they're hit by two precisely timed laser pulses. The first low-intensity laser pulse flattens the spherical droplet into a pancake shape, optimizing its surface area. Then, a second, much more powerful laser pulse vaporizes this flattened tin into a plasma that reaches 220,000 °C, 40 times hotter than the surface of the sun, which

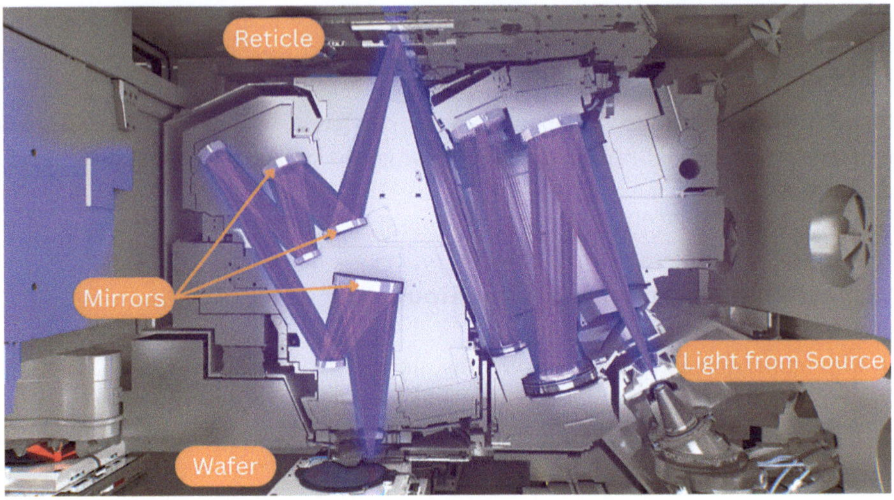


**Fig. 6.5** Simplified schematic of a EUV lithography system showing key components. Credit: ASML

emits the crucial EUV light. This two-step process happens 50,000 times per second to generate sufficient light for exposure.

Think of it like the world's most precise one-two punch. Imagine tiny droplets of liquid metal, each about 25 μm in diameter, being fired through the air at 70 m/s. The first laser pulse acts like a precision hammer, flattening each droplet into a thin pancake to maximize its surface area. Then, before the flattened tin can even begin to reform, a second, much more powerful laser delivers the knockout blow, instantly turning the metal into a glowing cloud of superheated plasma (Fig. 6.5).

After years of development, EUV lithography finally entered commercial production around 2019, starting with the 7 nm node at leading manufacturers. An EUV lithography machine is an engineering marvel, weighing over 150 tons, containing more than 100,000 parts, and costing upwards of $150 million each.

ASML's EUV tool is one of the most complex machines ever built, each of which costs over $150 million and contains precision optics so finely tuned that they can produce features just a few dozen atoms wide. The technology has advanced even further with newer ASML High-NA machines, costing upwards of $380 million, and is used for manufacturing chips with sub-3 nm processes.

The introduction of EUV lithography has enabled the continuation of Moore's Law, allowing manufacturers to create chips at the 5 nm node and below. While EUV is not used for every layer (it's primarily used for the most critical, fine-feature layers), it has become essential for cutting-edge semiconductor manufacturing.

Looking to the future, the industry is already working on the next generation of EUV tools with higher numerical aperture ("High-NA EUV"), which will enable even smaller features for future generations of chips at 3 nm and below.

### 6.4.3 Impact and Significance

The development of immersion lithography and EUV showcases the semiconductor industry's determination to overcome seemingly insurmountable challenges. These technologies have not only kept Moore's Law alive but also sustained the ongoing advancement of computing power that drives our modern digital world.

What's remarkable isn't just the engineering achievement but how these innovations in lithography have enabled everything from smartphones to artificial intelligence. Without these breakthroughs, digital progress would have stalled years ago.

The story of lithography innovation exemplifies the semiconductor industry's ethos: when faced with physical limits, find a way around them. This mindset has consistently delivered the exponential growth in computing capability that we've come to expect.

Now that we know how to pattern designs on the wafer, the next step is to start building those patterns in material. In Chap. 7, we will see how various materials, insulators, conductors, and semiconductors are added to the wafer, layer by layer (a process known as deposition).

**Summary**

- Photolithography is the process of transferring circuit patterns onto the silicon wafer using light. It's like printing tiny shapes using a photographic technique.
- A photomask (or reticle) carries the pattern for a chip layer. Light is shone through the mask onto the wafer. Where light hits the photoresist on the wafer, it changes the resist's solubility. After exposure, the wafer is developed to wash away either the exposed or unexposed resist, leaving a patterned resist that matches the mask pattern.
- Photoresist is a light-sensitive coating applied to wafers by spin-coating. Modern processes typically use positive resist, where exposed areas become soluble and are removed during development. This creates an etch mask on the wafer for subsequent steps.
- Photolithography equipment (steppers/scanners) use advanced optics to project the mask image onto the wafer with extreme precision, aligning each new pattern with previous layers. These tools historically used deep ultraviolet light (e.g., 193 nm wavelength excimer lasers) and have needed various tricks to push resolution.
- To continue shrinking feature sizes, engineers introduced immersion lithography (using a layer of water between lens and wafer to improve resolution at 193 nm) and eventually moved to EUV lithography at a 13.5 nm wavelength. EUV was a major technological leap, requiring new light sources and optics; after ~20 years of R&D, it's now used in cutting-edge fabs (enabling the latest nodes like 5 nm).
- Photolithography is a cornerstone of Moore's Law. Improvements in lithography have allowed the doubling of transistor density by enabling finer patterns on chips. Companies like ASML have been key players, developing the tools that make modern chip patterning possible.

## Review Questions

1. What is the role of a photomask in lithography, and how does it relate to the pattern that gets printed on the wafer?
2. Describe the basic steps of the photolithography process (coat, expose, develop). What happens to the photoresist during these steps?
3. Explain the difference between positive and negative photoresist. Which type results in the exposed areas being removed after development?
4. Why is the alignment of the mask pattern to the wafer important, and how do lithography tools ensure each layer lines up correctly with previous layers?
5. What was the purpose of introducing water immersion in 193 nm photolithography? How did it help improve resolution?
6. What is EUV lithography and why was it needed? List one major challenge that had to be overcome to make EUV lithography practical.

# 7 Deposition: Building Layers Atom by Atom

After learning about photolithography, the process that creates patterns on silicon, we now turn to another critical aspect of chip making, adding new materials to the wafer. This process, called deposition, is how we build up the intricate layers that form a functioning chip.

Think of semiconductor manufacturing as constructing a microscopic skyscraper. Photolithography provides the blueprint or floor plan, while deposition supplies the actual building materials, such as concrete, steel, and glass, that give the structure substance. In this chapter, we'll explore how manufacturers add these materials, often just atoms thick, with astonishing precision.

Deposition might sound straightforward, just adding material, but it encompasses a variety of sophisticated techniques tailored to specific materials and requirements. From growing insulating oxide layers to sputtering metal films for interconnections, these processes allow chip makers to build complex three-dimensional structures on a wafer's surface.

"The real magic of semiconductor manufacturing isn't just making things small," explains Dr. Karen, a veteran process engineer, "it's building these incredibly precise layer cakes of different materials, each layer serving a specific function and working in harmony with the others."

As we explore the world of deposition, we'll discover how these seemingly simple processes have enabled the entire digital revolution. After all, without these techniques to build up layers of materials, we would have no way to create the billions of transistors in modern chips.


A. D. Watpade, *From Sand to Chip*, Synthesis Lectures on Engineering, Science, and Technology, https://doi.org/10.1007/978-3-032-18439-9_7

## 7.1 Thermal Oxidation: Growing Silicon Dioxide Layers

Let's begin with one of the most fundamental deposition processes: thermal oxidation. This method exemplifies the elegant simplicity behind some of semiconductor manufacturing's most important techniques.

Imagine you're baking bread. As the dough heats in the oven, the outer surface transforms, developing a golden-brown crust with properties quite different from the soft interior. Something similar happens when we perform thermal oxidation on silicon.

Here's the recipe:

Take clean silicon wafers and place them in a specially designed furnace. Heat them to approximately 1000 °C (1830 °F), about five times hotter than your home oven's maximum temperature. Then introduce either pure oxygen gas (called "dry oxidation") or water vapor ("wet oxidation") into the furnace. As these silicon wafers "bake" in this oxygen-rich atmosphere, something remarkable happens.

The beautiful thing about silicon is that when exposed to oxygen at high temperatures, it reacts to form silicon dioxide, essentially a high-quality glass, right on its surface (Fig. 7.1).

Unlike our bread analogy, this isn't just a surface color change. The oxygen actually combines with the silicon atoms, transforming the uppermost layer of the wafer into silicon dioxide ($SiO_2$). This is somewhat like how iron forms rust when exposed to oxygen, except that while rust is typically undesirable, this "silicon rust" is precisely what we want.

The oxide layer grows directly from the silicon, consuming some of the original wafer in the process. For every 0.44 μm of silicon consumed, approximately 1 μm of oxide forms, and the layer expands as oxygen atoms integrate into the silicon crystal structure.

Control is the key advantage here. By adjusting the temperature and time, engineers can grow oxide layers with precisely controlled thickness, from just a few nanometers (a few dozen atoms thick) to over a micrometer.

"Silicon dioxide was the foundation that made integrated circuits possible," explains Dr. Tsu Liu, noting the historical significance. Indeed, in 1959, Jean Hoerni at Fairchild

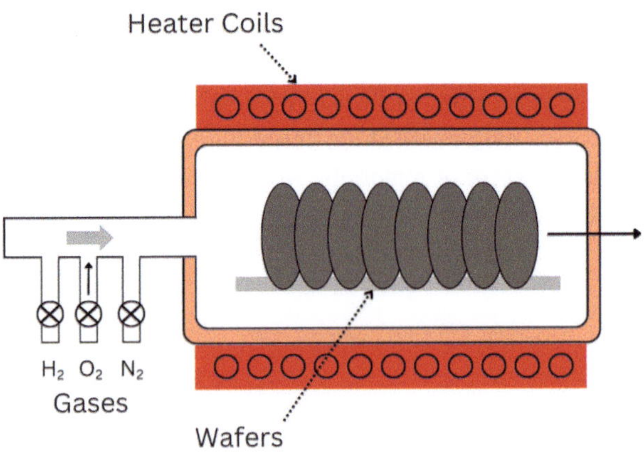


**Fig. 7.1** Thermal oxidation process in a diffusion furnace, showing silicon wafers "baking" in oxygen, forming a silicon dioxide layer much like bread forms a crust in an oven

Semiconductor invented the "planar process," using silicon dioxide layers to protect transistors, which revolutionized the industry and made modern chips possible.

This silicon dioxide layer serves several crucial functions:

- It acts as an **excellent electrical insulator**, preventing short circuits between components
- It creates a **protective layer** that shields delicate silicon from contamination
- It forms a **selective barrier** for subsequent processing steps (like doping, which we'll cover in a later chapter)
- It provides **electrical stability** for transistors by reducing surface defects

Once grown, this oxide layer becomes a versatile tool in the chipmaker's arsenal. Areas of the oxide can be selectively removed (using the photolithography and etching techniques we've discussed) to create windows for introducing dopants or forming electrical connections. The remaining oxide protects and insulates other regions.

This $SiO_2$ might serve as the gate dielectric in a transistor or a protective layer during doping processes. In many ways, this oxide is the foundation upon which modern transistors are built.

The thermal oxide's quality is exceptional because it grows from the silicon itself, forming an almost perfect interface with minimal defects. This interface quality is essential for reliable transistor operation.

As we progress in our chipmaking journey, this silicon dioxide layer will become the foundation for many subsequent steps, a perfect example of how semiconductor manufacturing carefully builds complexity one layer at a time.

## 7.2 Chemical Vapor Deposition (CVD): Laying Down Films from Gas

While thermal oxidation is excellent for growing silicon dioxide, chip manufacturers need to deposit many other materials as well. Chemical Vapor Deposition (CVD) is a remarkably versatile technique that enables the deposition of a wide range of films.

Imagine you're walking outdoors on a cold winter morning. Your warm breath condenses in the air, forming visible vapor. If you breathe onto a cold window, tiny water droplets form on the glass surface. Chemical Vapor Deposition works on a similar principle, but with far more precision and using reactive gases rather than water vapor.

Here's how CVD works: Wafers are placed in a special chamber and heated. Then, carefully selected gases are introduced. When these gases touch the hot wafer surface, they react or decompose, forming a solid material that deposits as a thin, uniform film.

It's like magic watching it happen. You start with nothing but gases, and within minutes, an incredibly uniform layer of material appears on the wafer, built one atom at a time (Fig. 7.2).

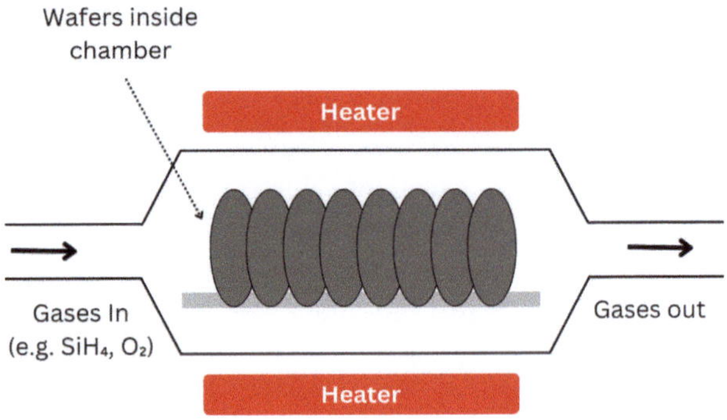


**Fig. 7.2** Diagram of a CVD chamber showing gas flow, reaction at the wafer surface, and film formation

Unlike thermal oxidation, which converts the silicon itself, CVD adds entirely new material to the wafer surface. The gases used contain the elements needed for the film. For example:

- Silane ($SiH_4$) provides silicon atoms
- Ammonia ($NH_3$) or nitrogen ($N_2$) supplies nitrogen
- Various other gases contribute oxygen, carbon, or metals as needed

When these gases react at the wafer surface, the desired elements bond together while gaseous byproducts are pumped away. The result is a solid film with precisely controlled composition.

The versatility of CVD is remarkable. It can deposit:

- Insulating materials like silicon dioxide ($SiO_2$) and silicon nitride ($Si_3N_4$)
- Semiconducting layers like polycrystalline silicon ("polysilicon")
- Conducting materials for certain applications
- Special compound semiconductors for advanced devices

For example, CVD might deposit an insulating silicon nitride layer to isolate transistors from each other, or a doped polysilicon layer that could act as a transistor gate.

What's amazing about CVD is how precisely we can control the film properties by adjusting the gas mixture, pressure, and temperature. We can fine-tune the density, stress, electrical properties, and even create gradients of composition if needed.

Several CVD variants exist for specific applications:

- **Low-Pressure CVD (LPCVD)**: Operates at reduced pressure (hence the name) and higher temperatures (around 600–800 °C). It's excellent for depositing very uniform films across many wafers simultaneously, making it ideal for polysilicon and silicon nitride layers. The furnaces look similar to oxidation furnaces, with wafers arranged vertically in quartz boats.

- **Plasma-Enhanced CVD (PECVD)**: Uses plasma (an energized gas with charged particles) to enhance the chemical reactions. This allows deposition at lower temperatures (around 300–400 °C), preventing damage to existing structures on the wafer. PECVD is particularly useful for depositing insulating layers later in the process when high temperatures might damage metal connections.

For ultra-thin films with atomic-level precision, an advanced technique called **Atomic Layer Deposition (ALD)** has been developed. ALD works by introducing gases one at a time in sequential pulses, building the film literally one atomic layer per cycle, like painting with atomic precision, one coat at a time.

CVD became crucial in the 1970s and 1980s as chips grew more complex, requiring new materials beyond what oxidation could provide. For instance, silicon nitride layers deposited by CVD became vital as a barrier against moisture and sodium contamination, which could ruin transistor reliability.

Today, CVD is an indispensable part of the chipmaking process, used repeatedly to build the complex stack of materials that form modern integrated circuits. From the gate oxides in transistors to the insulating layers between metal interconnections, CVD-deposited films are everywhere in your smartphone or computer.

## 7.3 Epitaxy: Growing Perfect Silicon Layers on Silicon

Epitaxy is perhaps the most elegant of all deposition techniques, a process that seems almost magical in its precision. The word "epitaxy" comes from Greek, meaning "arranged upon," which perfectly describes what happens: a new crystal grows upon an existing crystal, extending its ordered structure.

Imagine you have a perfectly cut gemstone, a large diamond with its carbon atoms precisely arranged in a crystal lattice. Now imagine if you could somehow add more carbon atoms to it, causing the diamond to grow larger while maintaining that perfect crystalline structure. That's essentially what epitaxy accomplishes with silicon.

Epitaxy is like building the perfect extension to a house. The new structure perfectly matches and continues the pattern of the original, so you can't tell where one ends and the other begins.

In an epitaxial process, silicon wafers are placed in a special reactor and heated to temperatures around 1000–1200 °C. Then, a silicon-containing gas (typically silane or silicon tetrachloride) flows over the hot wafer surface. Under these controlled conditions, silicon atoms from the gas deposit onto the wafer surface, arranging themselves to perfectly continue the crystal structure of the underlying silicon (Fig. 7.3).

What makes epitaxy remarkable is that unlike typical CVD processes (which often create polycrystalline or amorphous films), epitaxial deposition creates a single-crystal film that is structurally indistinguishable from the original wafer, except for one important aspect that it can have different properties.

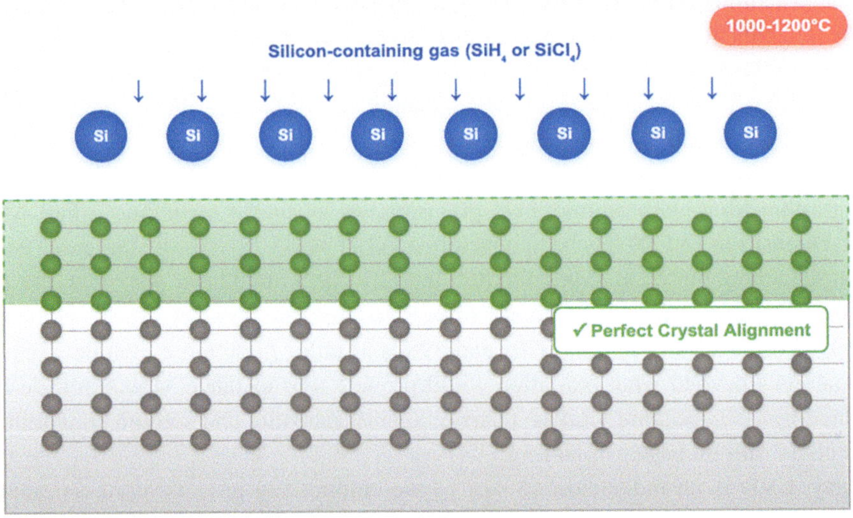


**Fig. 7.3** Epitaxial growth showing silicon atoms arranging themselves to continue the crystal pattern of the substrate

By adding small amounts of other elements (dopants) to the gas flow during epitaxy, engineers can precisely control the electrical characteristics of the grown layer. For example, they might grow a lightly doped epitaxial layer on top of a heavily doped silicon substrate, creating an ideal foundation for building transistors with specific electrical properties.

The beauty of epitaxy is that it gives us the 'best of both worlds'. We can start with one type of silicon and add a perfect layer of different silicon on top, tailoring the wafer to our exact needs.

Common applications of epitaxy include:

- Growing a lightly doped layer on a heavily doped substrate for better transistor isolation
- Creating buried layers with special electrical properties
- Forming specialized structures for power devices
- Growing compound semiconductor layers for high-performance applications
- Creating silicon-germanium layers for advanced transistors

Manufacturers often grow an epitaxial silicon layer to get a very pure, controlled starting surface for transistors after the initial wafer preparation.

Historically, epitaxy became crucial in the 1980s as device designs grew more sophisticated. For example, bipolar transistors benefited enormously from epitaxial layers that improved their speed and reliability. Later, as CMOS technology (the basis for most modern chips) advanced, epitaxy provided the precise control needed for ever-smaller transistors.

Modern variations of epitaxy can even grow different semiconductor materials on silicon, such as silicon-germanium or III-V compounds, opening new possibilities for high-performance devices.

The ability to grow perfect crystal layers with precisely controlled properties epitomizes the atomic-level precision that makes modern semiconductor manufacturing so remarkable. While most people never see or think about epitaxy, this process has been essential to the advancement of the chips that power our digital world.

## 7.4 Physical Vapor Deposition: Sputtering Metal and More

So far, we've explored methods that rely on chemical reactions to create layers on our wafer. But what about metals like aluminum or copper that form the vital connections between transistors? For these materials, we turn to a different approach: Physical Vapor Deposition (PVD), with sputtering being the most common technique.

CVD and epitaxy are chemical methods that build material via chemical reactions, whereas sputtering is a physical method that works by knocking atoms off a source material. This fundamental difference makes each technique suited for different materials and applications.

Imagine you have a plate of metal and you shoot tiny bullets at it, atoms get knocked off the plate and fly through space, eventually landing on and coating nearby surfaces. That's essentially how sputtering works, though the "bullets" are actually ions, charged atoms in a plasma.

"It's a bit like atomic billiards," explains Dr. John, a PVD expert. "We knock atoms off a target material, and they travel across the chamber to coat our wafers."

Here's how the sputtering process works:

First, wafers are placed in a vacuum chamber opposite a "target" made of the material we want to deposit, typically a pure metal like aluminum, titanium, or copper. The chamber is filled with a small amount of an inert gas, usually argon.

Then, a high voltage is applied, creating a plasma, a gas of electrically charged particles. The positively charged argon ions in this plasma are accelerated toward the negatively charged target with tremendous energy.

When these ions strike the target, they knock metal atoms loose, like a cue ball striking a rack in billiards. These dislodged metal atoms fly away from the target in all directions. Many land on the wafer, building up a thin metal film atom by atom (Fig. 7.4).

Unlike CVD, which relies on chemical reactions, sputtering is primarily a physical process, atoms are mechanically knocked off the target and physically travel to the wafer. This makes sputtering ideal for materials that don't readily form volatile compounds needed for CVD.

"What's fascinating about sputtering," notes Dr. Lisa Chen, a metallization expert, "is that we're literally building metal films one atom at a time, yet we can coat an entire 300 mm wafer with a perfectly uniform layer in minutes."

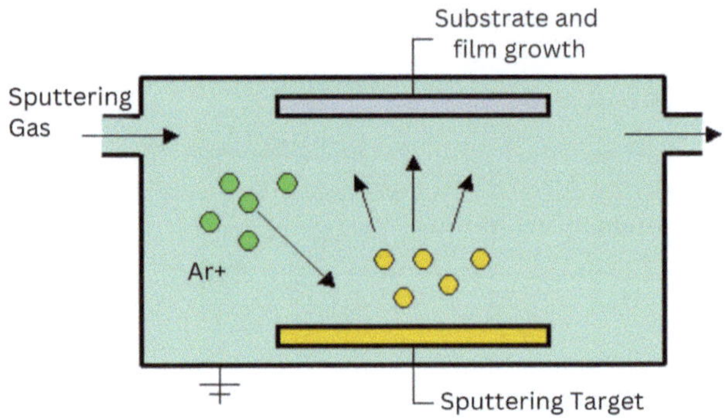


**Fig. 7.4** Sputtering process showing argon ions striking a metal target and dislodged atoms coating the wafer

Sputtering is used to deposit various materials:

- Metal layers for circuit connections (aluminum, copper seed layers)
- Barrier metals (titanium, tantalum) that prevent unwanted reactions
- Special alloys for specific electrical properties
- Some non-metallic materials like certain oxides

Sputtering is widely used to deposit the metal wiring and other metal films (like adhesion or barrier layers) in a chip.

A related PVD technique is thermal evaporation, where the source material is simply heated until it evaporates or sublimates, and its atoms travel through vacuum to coat the wafer. While conceptually simpler than sputtering, evaporation doesn't work as well for materials with high melting points and typically produces less uniform films.

Sputtering became crucial to the semiconductor industry as chips grew more complex. Early integrated circuits in the 1960s used aluminum connections deposited by evaporation, but as features shrank and wafers grew larger, sputtering took over due to its better uniformity and control.

Modern sputtering systems are remarkably sophisticated, with multiple chambers that can deposit several different materials in sequence without breaking vacuum. This allows the formation of complex metal stacks needed for advanced chips.

Every computer chip in your devices contains metal layers deposited by sputtering, the aluminum or copper "wires" that connect billions of transistors together. Without this high-tech atomic spray painting, our advanced electronic devices simply couldn't function.

Now that we've learned how to add layers of various materials onto the wafer, the next step is removing selected parts of these layers to form patterns, a process called etching. In Chap. 8, we'll see how both liquid chemicals and plasmas are used to carve those deposited films according to the patterns from photolithography.

## Summary

- Deposition processes add new materials to the silicon wafer, building the chip layer by layer. These techniques allow precise control of film thickness and properties, from a few atoms thick to several micrometers.
- Thermal oxidation grows silicon dioxide (essentially glass) directly from the silicon wafer by heating it in oxygen or steam. This oxide serves as an excellent insulator, protective layer, and masking material for subsequent steps.
- Chemical Vapor Deposition (CVD) creates films by flowing reactive gases over the heated wafer surface, causing them to react or decompose and form a solid layer. CVD can deposit a wide variety of materials, including insulators like silicon nitride, semiconductors like polysilicon, and specialized films.
- Epitaxy grows perfect single-crystal silicon layers that extend the wafer's crystal structure. Unlike other depositions, epitaxial layers maintain atomic order while allowing controlled changes in properties like doping concentration.
- Physical Vapor Deposition (PVD), particularly sputtering, deposits metals and other materials by physically dislodging atoms from a target and allowing them to coat the wafer. This technique is essential for creating the metal interconnections that link transistors together.

## Review Questions

1. What is the purpose of deposition in semiconductor manufacturing? Give two examples of materials that might be deposited on a silicon wafer and why they are needed in a chip.
2. How is thermal oxidation of silicon performed, and what material is produced on the wafer surface? Why is this oxide layer important for chip fabrication?
3. Describe in simple terms how chemical vapor deposition (CVD) creates a thin film on a wafer. What is the role of the gases and the wafer's temperature in this process?
4. Explain what epitaxy is. How is an epitaxial silicon layer different from a polysilicon layer deposited by standard CVD?
5. What is sputtering, and how does it differ from CVD? Name a common application of sputtering in making a microchip (what kind of material is typically deposited this way).
6. Why do you think chip makers use several different deposition techniques (oxidation, CVD, PVD, etc.) rather than just one? (Think about factors like types of materials, temperatures, and film qualities.)
7. (Research exercise) Silicon dioxide can be deposited by CVD or grown by oxidation. Find out one advantage of each method (grown oxide vs. deposited oxide) and in what situation each might be used.

# 8 Etching: Carving Patterns into Silicon

Picture a sculptor, carefully chiseling away stone to reveal the masterpiece hidden within. In semiconductor manufacturing, etching plays a similar role as it is the process where we selectively remove material to create the intricate patterns that make up a microchip. After we have deposited various materials on the wafer and used photolithography to create a pattern template, etching is what actually transfers these patterns into the underlying layers.

Etching might sound destructive at first, after all, we are removing material rather than adding it. But this controlled removal is absolutely essential to creating the complex, three-dimensional structures that allow modern chips to function. Without precise etching, the billions of transistors and interconnections in today's chips simply wouldn't be possible.

In this chapter, we'll explore the two main approaches to etching, wet etching, which uses liquid chemicals, and dry etching, which uses plasma, and see how semiconductor manufacturers choose the right technique for each step in the manufacturing process (Fig. 8.1).

## 8.1 Wet Etching: Dissolving Material with Chemicals

Imagine you're creating a custom metal necklace. You might apply a protective coating in certain areas, then dip the metal into an acid bath that dissolves the unprotected portions, leaving behind your desired pattern. This is essentially how wet etching works in semiconductor manufacturing, albeit at a much smaller scale and with far more precision.



A. D. Watpade, *From Sand to Chip*, Synthesis Lectures on Engineering, Science, and Technology, https://doi.org/10.1007/978-3-032-18439-9_8

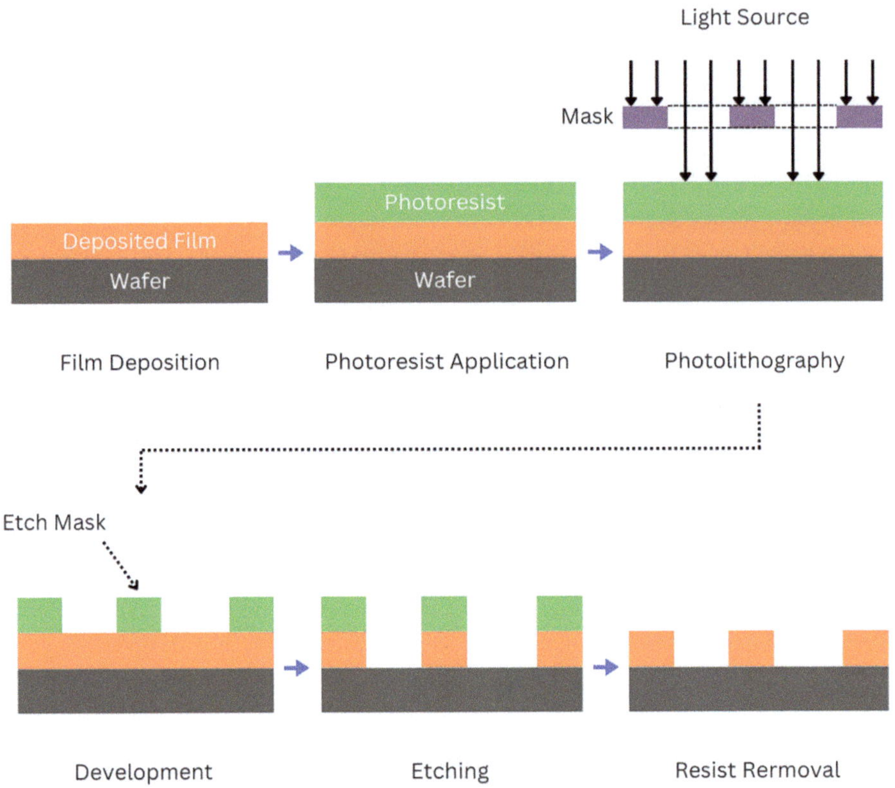


**Fig. 8.1** Pattern Transfer Process: From Film Deposition to Final Structure. This sequence shows how patterns are created in semiconductor films. The process is the same no matter which etching method is used in step 5 and described further in this chapter

### 8.1.1 How Wet Etching Works

Wet etching uses liquid chemicals to dissolve unwanted material from the wafer surface. The process is relatively straightforward:

1. Start with a wafer that has a layer of material we want to pattern (like silicon dioxide)
2. Apply photoresist and use photolithography to create a protective pattern
3. Immerse the wafer in a chemical solution (Etch bath) that dissolves the exposed material
4. Rinse and dry the wafer
5. Remove the remaining photoresist

The chemical solution (the "etchant") dissolves the target material through a chemical reaction, typically converting solid material into soluble compounds that simply wash away in the liquid (Fig. 8.2).

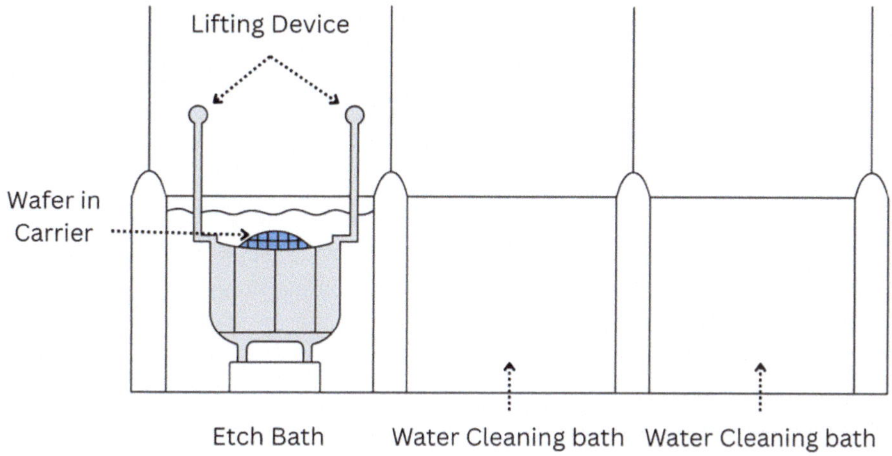


**Fig. 8.2** Wet Etching Process showing a wafer with photoresist pattern immersed in an etchant solution

Think about etching as similar to how people make traditional printed circuit boards at home. A copper-clad board is covered with a protective pattern, then immersed in a solution that dissolves exposed copper but doesn't affect the protected areas. After enough time, the exposed copper disappears, leaving behind only the protected circuit traces.

### 8.1.2 Selectivity: Choosing the Right Chemical for the Job

One of the most important concepts in etching is selectivity, the ability of an etchant to remove one material while leaving others intact. It's like having a magic eraser that removes only pencil marks but leaves ink untouched.

For example, hydrofluoric acid (HF) is commonly used to etch silicon dioxide (glass) because it dissolves the oxide relatively quickly while attacking silicon and most photoresists much more slowly. This selectivity is crucial because we want to remove the target material without damaging the pattern or the underlying layers.

Some common wet etchants include:

- Hydrofluoric acid (HF) for silicon dioxide
- Phosphoric acid for aluminum
- Potassium hydroxide (KOH) for silicon

### 8.1.3 Isotropic Etching: The Sideways Challenge

A key characteristic of most wet etching processes is that they are isotropic, meaning they etch in all directions at roughly the same rate. Imagine dropping a sugar cube in water, as it dissolves from all sides equally. Similarly, most wet etchants do not just etch downward, but also sideways under the protective photoresist.

This sideways etching, called "undercutting," creates a rounded profile beneath the edges of the resist pattern. For larger features, this slight undercutting might be acceptable. But as chip features have shrunk to microscopic dimensions, this undercutting has become increasingly problematic (Fig. 8.3).

Consider a simple example. If you're trying to etch a trench that's 1 µm wide, and your etchant undercuts by 0.5 µm on each side, your final trench width will be 2 µm, which is twice what you intended. This is why wet etching, despite its simplicity, isn't suitable for the smallest features in modern chips.

### 8.1.4 Crystal-Dependent Etching

Interestingly, some wet etchants show anisotropic behavior when etching crystalline materials like silicon. For example, potassium hydroxide (KOH) etches silicon at different rates depending on the crystal plane orientation. It etches much faster along some crystal planes than others, allowing for V-shaped grooves or pyramidal pits with precise angles.

This property is particularly useful in creating microelectromechanical systems (MEMS) devices like pressure sensors or inkjet printer nozzles, where these angled structures are beneficial.

### 8.1.5 Historical Perspective

In the early days of integrated circuit manufacturing (1960s–1970s), wet etching was the primary method used. Technicians in bunny suits would manually dip wafers into acid

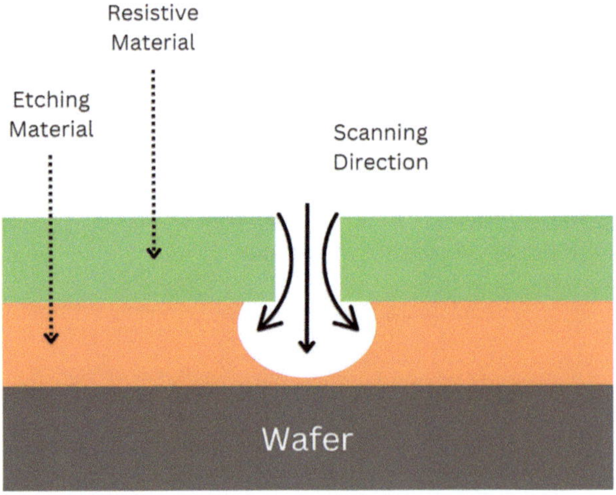


**Fig. 8.3** Isotropic Wet Etching and Undercutting showing how etching occurs sideways under the resist mask

baths and carefully time the process. One misstep, such as leaving the wafer in too long or using the wrong concentration, could ruin an entire batch.

As feature sizes shrank below a few microns in the late 1970s and early 1980s, the limitations of wet etching became increasingly apparent. The industry needed more precise methods that could create smaller, more vertical features. This need drove the development and adoption of dry etching techniques.

### 8.1.6 Modern Applications of Wet Etching

Despite its limitations for the smallest features, wet etching still has important applications in modern semiconductor manufacturing:

- Cleaning and surface preparation
- Etching larger features where undercutting isn't critical
- Removing damaged layers
- Creating certain MEMS structures using anisotropic crystalline etching
- Situations where selectivity is the primary concern

Wet etching remains valuable because it's relatively simple, offers excellent selectivity for certain material combinations, and can be very cost-effective for appropriate applications.

## 8.2 Dry (Plasma) Etching: Precision Carving with Ions

While wet etching served the semiconductor industry well in its early days, the drive toward smaller, more densely packed circuits required more precise etching methods. Enter dry etching, a revolutionary approach that uses ionized gas (plasma) instead of liquid chemicals to remove material. This technological leap was a major enabler of Moore's Law, allowing feature sizes to continue shrinking well beyond what wet etching could achieve.

### 8.2.1 Understanding Plasma

Before diving into dry etching, let's quickly understand what plasma is. Often called the "fourth state of matter" (beyond solid, liquid, and gas), plasma is a gas where some atoms have been stripped of their electrons, creating a mixture of ions (positively charged) and free electrons (negatively charged). Despite being electrically charged, the overall plasma remains electrically neutral.

A plasma glows, much like a neon sign, because of energy released when electrons recombine with ions. In semiconductor manufacturing, we create plasma by applying strong electromagnetic fields to gases at low pressure.

### 8.2.2 How Dry Etching Works

Imagine a microscopic sandblaster combined with a targeted chemical reaction. That's essentially how dry etching works. Here's a simplified version of the process (Fig. 8.4).

1. The wafer with its photoresist pattern is placed in a vacuum chamber
2. A gas is introduced and converted to plasma using electromagnetic energy
3. The plasma creates reactive species (ions and radicals)
4. These reactive species interact with the exposed surface, removing material
5. The etching byproducts (now gaseous) are pumped away

### 8.2.3 Types of Dry Etching

Dry etching works through two main mechanisms, often operating simultaneously:

- **Physical Etching (Sputtering):** Like a microscopic sandblaster, energetic ions from the plasma are accelerated toward the wafer surface. When they strike the surface, they physically knock atoms loose through momentum transfer. This process is highly directional since the ions mostly travel straight down toward the wafer.
- **Chemical Etching:** Reactive species in the plasma (like fluorine or chlorine radicals) chemically react with the surface material, forming volatile compounds that desorb from the surface and get pumped away. For example, fluorine radicals react with silicon to form silicon tetrafluoride ($SiF_4$), a gas that easily leaves the surface.

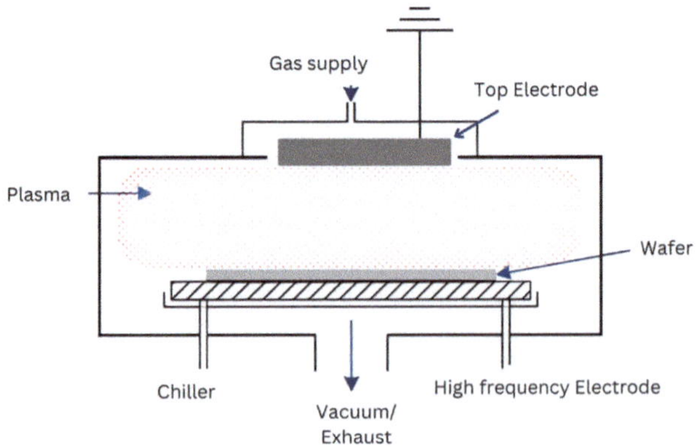


**Fig. 8.4** Reactive Ion Etching (RIE) chamber showing electrodes, plasma, wafer, and gas flow. The RF power creates plasma between the electrodes, and ions accelerate toward the wafer

Most practical dry etching systems use both mechanisms, a combination known as Reactive Ion Etching (RIE). The physical component provides directionality, while the chemical component enhances the etch rate and selectivity.

### 8.2.4 Anisotropic Etching: Going Straight Down

The biggest advantage of dry etching is its ability to create anisotropic profiles, etching primarily downward with minimal sideways etching. This is possible because

1. The ions in the plasma are accelerated perpendicular to the wafer surface
2. The physical sputtering component occurs primarily in this downward direction
3. In many processes, protective polymer films form on sidewalls during etching, preventing lateral etching

This anisotropy allows for much smaller, more tightly packed features with nearly vertical sidewalls, which is essential for modern high-density chips (Fig. 8.5).

### 8.2.5 Dry Etching Equipment

A basic plasma etching system includes:

- A vacuum chamber with controlled gas flow
- Electrodes to create and shape the plasma (various configurations exist)
- A power source (typically radio frequency)
- A wafer holder (often temperature-controlled)
- Pumps, gauges, and control systems

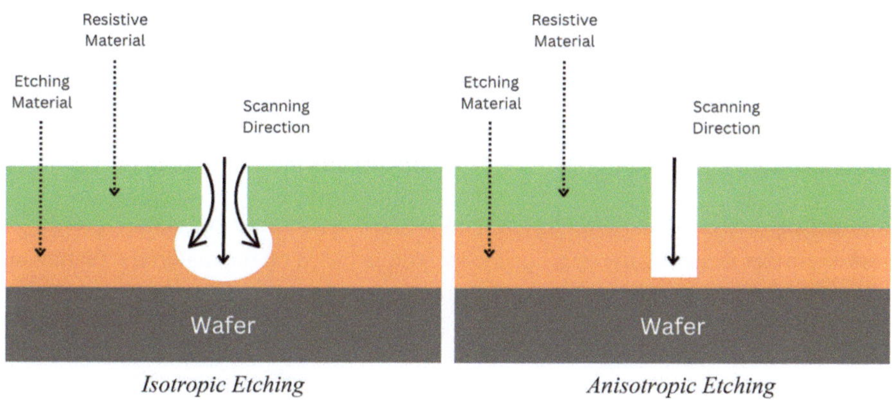


**Fig. 8.5** Comparison of wet etching (isotropic with undercutting) versus dry etching (anisotropic with vertical walls)

More advanced systems may add features like magnetic enhancement, multiple power sources, or endpoint detection systems that can tell when the etching has reached the desired depth.

It's worth noting that these plasma processes use gases like $CF_4$ (carbon tetrafluoride) or $Cl_2$ (chlorine), the same toxic gases we discussed in Chap. 5's safety systems. This is why semiconductor fabs need those extensive gas detection systems, emergency shutoffs, and specialized scrubbers to treat exhaust gases before they leave the facility.

### 8.2.6 The Game-Changer for Moore's Law

The introduction of plasma etching in the late 1970s and early 1980s was revolutionary for the semiconductor industry. With wet etching alone, feature sizes would have hit a wall around 1–2 μm due to undercutting problems. Plasma etching, with its ability to create vertical profiles, helped enable the continued miniaturization of transistors.

When engineers at companies like Bell Labs and Intel first demonstrated reliable plasma etching of sub-micron features with straight sidewalls, it opened the door to generations of smaller, faster, and more powerful chips. Today's advanced chips, with features measured in nanometers rather than microns, simply wouldn't be possible without dry etching technology.

### 8.2.7 Types of Dry Etching Processes

Several variations of dry etching are used in modern semiconductor manufacturing:

- **Reactive Ion Etching (RIE):** This is the most common type of etching process where wafers are placed on an electrode that creates both the plasma and an electrical field that accelerates ions toward the wafer.
- **Inductively Coupled Plasma (ICP) Etching:** Uses a coil to generate a high-density plasma, allowing for faster etch rates and better control.
- **Deep Reactive Ion Etching (DRIE):** A specialized technique for creating extremely deep, high-aspect-ratio structures by alternating between etching and protective polymer deposition (like the Bosch process for MEMS devices).

Each of these processes has found its niche in modern semiconductor manufacturing, with process engineers selecting the appropriate technique based on the materials, feature sizes, and requirements of each layer.

## 8.3 Etch Profiles and Selectivity: Getting the Right Shape Without Damage

Creating a good etch isn't just about removing material. It's about removing exactly the right material, in exactly the right shape, stopping at exactly the right time. Two critical concepts govern this precision: etch profiles and selectivity.

### 8.3.1 Etch Profiles: The Shape Matters

The sidewall profile, the shape of the "walls" created by etching, is crucially important in semiconductor manufacturing (Fig. 8.6). Different profiles include:

- **Vertical (Anisotropic):** Straight sidewalls with minimal undercutting, crucial for densely packed features. Most advanced logic and memory devices require vertical profiles.
- **Tapered:** Sidewalls that angle outward or inward slightly, which can be beneficial for certain applications like creating vias (vertical connections between layers) where subsequent deposition steps need to fill the etched hole.
- **Undercut (Isotropic):** Rounded profiles with significant etching beneath the mask, typically seen in wet etching. While problematic for tiny features, this can sometimes be useful for creating specific structures like air bridges or for releasing MEMS structures.

The profile needs to match the intended function of the structure. For example, a transistor gate typically needs vertical sidewalls to ensure consistent electrical properties, while a contact hole might benefit from slightly tapered walls for better metal filling.

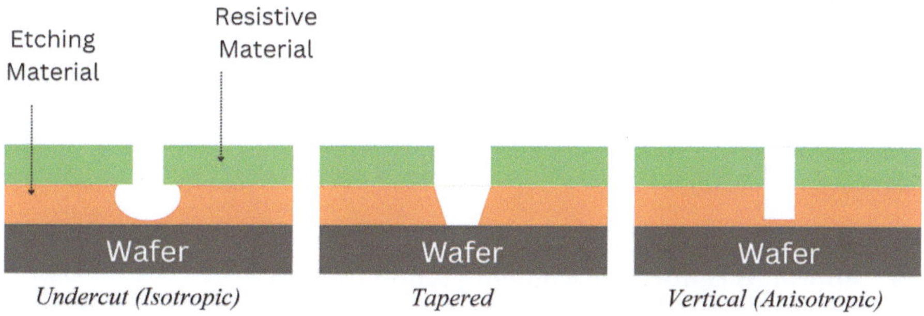


**Fig. 8.6** Common Etch Profiles showing vertical (anisotropic), tapered, and undercut (isotropic) shapes

### 8.3.2 Controlling the Profile

In dry etching, engineers can control the profile by adjusting several parameters:

- Gas composition: Different gases create different balances between physical and chemical etching
- Pressure: Lower pressure generally gives more directional, anisotropic etching
- Power: Higher power increases ion energy and enhances the physical component
- Temperature: Affects chemical reaction rates and polymer formation
- Bias voltage: Controls ion acceleration toward the wafer

By fine-tuning these parameters, process engineers can achieve the exact profile needed for each layer of the chip.

### 8.3.3 Selectivity: Etching What You Want, Protecting What You Don't

Selectivity is the ratio of how fast the target material etches compared to other materials present. High selectivity means the etch removes the target material much faster than others.

For example, if we're etching silicon dioxide that sits on top of silicon, we want an etch process with high oxide-to-silicon selectivity. It should remove the oxide rapidly but barely touch the silicon underneath. Similarly, we want good selectivity to the photoresist mask so that it remains intact throughout the etch process.

Think of selectivity like using sandpaper that quickly removes paint but barely scratches the wood underneath. Without good selectivity, we might damage underlying layers or lose our pattern before the etch is complete.

### 8.3.4 The Balancing Act

Process engineers often face trade-offs between profile and selectivity. A process that gives perfectly vertical sidewalls might have lower selectivity, while a highly selective process might not achieve the ideal profile. Finding the right balance for each specific application is part science, part art.

### 8.3.5 Endpoint Detection: Knowing When to Stop

Another critical aspect of etching is knowing when to stop. Etch too little, and residual material may remain where it should be cleared. Etch too much, and you might damage underlying layers or create features with the wrong dimensions.

Modern etching systems use various endpoint detection methods:

- Optical emission spectroscopy: Monitoring light emitted by the plasma, which changes when the composition of the etched material changes
- Laser interferometry: Measuring the thickness of the etching layer in real-time
- Mass spectrometry: Analyzing the composition of gases exhausted from the chamber

These techniques allow for precise control over etch depth and help ensure consistency across wafers and batches.

### 8.3.6 Real-World Challenges

As feature sizes have continued to shrink, etching has faced increasing challenges:

- **Aspect Ratio Dependent Etching (ARDE):** Narrow, deep features etch more slowly than wider features, making it difficult to etch all features to the same depth simultaneously.
- **Microloading:** Areas with dense patterns etch differently than isolated features, creating non-uniformity across the wafer.
- **Pattern Dependency:** The local pattern density affects etch rates and profiles, requiring complex compensation techniques.
- **Critical Dimension (CD) Control:** Maintaining precise feature dimensions throughout the etch process becomes increasingly difficult at smaller nodes.

To overcome these challenges, advanced etching processes may use multi-step approaches, specialized gas chemistries, pulsed plasmas, or other techniques developed through decades of research and engineering.

### 8.3.7 The Impact on Device Performance

The quality of the etch directly impacts device performance. Even small variations in profile, critical dimensions, or surface damage can affect transistor speed, power consumption, or reliability. As chips have become more complex and feature sizes have shrunk, the precision requirements for etching have become increasingly stringent.

For example, in modern FinFET transistors, the exact profile and smoothness of the etched fin structure significantly impacts the transistor's electrical characteristics. A slight variation in the fin width or profile can change how the transistor performs, potentially affecting the entire chip.

This is why leading-edge semiconductor manufacturers invest heavily in advanced etching equipment and process development. The precision of the etch can make or break a chip design.

We can now build up layers and etch them into circuits. But one critical step remains to turn our patterned silicon into active devices: doping. In Chap. 9, we'll see how we introduce specific impurities into selected regions of the silicon to create transistors and other components, essentially putting the "semi" in semiconductor.

## Summary

- Etching is the process of selectively removing material to create patterns in semiconductor manufacturing, working hand-in-hand with photolithography to transfer patterns onto the wafer.
- Wet etching uses liquid chemicals to dissolve material and is typically isotropic (etches in all directions equally), causing undercutting that limits its use for the smallest features in modern chips.
- Dry (plasma) etching uses ionized gases to remove material through physical bombardment and chemical reactions, enabling anisotropic (directional) etching with vertical sidewalls that's essential for today's advanced chips.
- Etch profile (the shape of the sidewalls) and selectivity (preferentially removing one material while preserving others) are crucial parameters that engineers must carefully control to create functioning semiconductor devices.
- The transition from wet to dry etching in the 1970s–1980s was a key enabler of continued miniaturization in line with Moore's Law, allowing for the creation of smaller transistors and more powerful chips.

## Review Questions

1. What is the main purpose of the etching step in chip manufacturing? How does it relate to photolithography and deposition in the process of creating patterns?
2. Explain the difference between wet etching and dry etching in your own words. What does each use to remove material, and what is one advantage of dry etching for very small features?
3. What does it mean for an etch to be "anisotropic" versus "isotropic"? Which type of etching (wet or plasma) is more anisotropic, and why is that important for modern chips?
4. Define "selectivity" in the context of etching. Why is selectivity important when etching a layer that sits on top of other materials or is protected by photoresist?
5. During a wet etch, what can happen to the sides of the features under the photoresist, and why might that be a problem as circuits get smaller?
6. Dry etching often involves a plasma. What role do the ions in a plasma etch play in removing material from the wafer?

# Doping: Putting the "Semi" in Semiconductor

# 9

In our journey through semiconductor manufacturing, we've explored how silicon wafers are prepared, patterned with photolithography, enhanced with deposited films, and carved with etching. Yet there's a crucial step that truly makes silicon work as an electronic material, and that is **doping**.

**Doping** is the process that transforms ordinary silicon into a functional semiconductor by adding tiny amounts of carefully selected impurities. Without doping, silicon would remain almost an insulator, and none of our modern electronic devices would work. It's literally what puts the "semi" in semiconductor!

In this chapter, we'll discover why doping is essential, how it creates the foundation for transistors and other electronic components, and the fascinating methods manufacturers use to precisely place dopant atoms where they're needed.

## 9.1 Why Doping Is Needed: Creating P-N Junctions

Pure silicon, despite being the backbone of the semiconductor industry, isn't particularly useful on its own. As we learned in Chap. 3, pure silicon at room temperature is almost an insulator because it doesn't conduct electricity well at all. This is because in pure silicon, the electrons are tightly bound in perfect crystalline bonds, leaving few free carriers to conduct current.

### 9.1.1 The Magic Pinch of Impurity

Doping is like adding a secret ingredient to a recipe. Just a tiny pinch changes everything. By introducing specific impurity atoms into the silicon crystal, we can dramatically alter

 
A. D. Watpade, *From Sand to Chip*, Synthesis Lectures on Engineering, Science, and Technology, https://doi.org/10.1007/978-3-032-18439-9_9

its electrical behavior. These impurity atoms are called **dopants** and they come in two main varieties (Fig. 9.1).

- **N-type dopants** include elements like phosphorus, arsenic, or antimony, which have five valence electrons (compared to silicon's four). When incorporated into the silicon lattice, these extra electrons aren't needed for bonding and can move freely through the material, enhancing electrical conductivity. The "n" stands for negative, referring to the negatively charged electron carriers.
- **P-type dopants** include elements like boron, gallium, or indium, which have three valence electrons. When these atoms replace silicon in the crystal lattice, they create a hole, a missing electron in the bonding structure. These holes act as positive charge carriers that can move through the material. The "p" stands for positive, referring to the positively charged hole carriers.

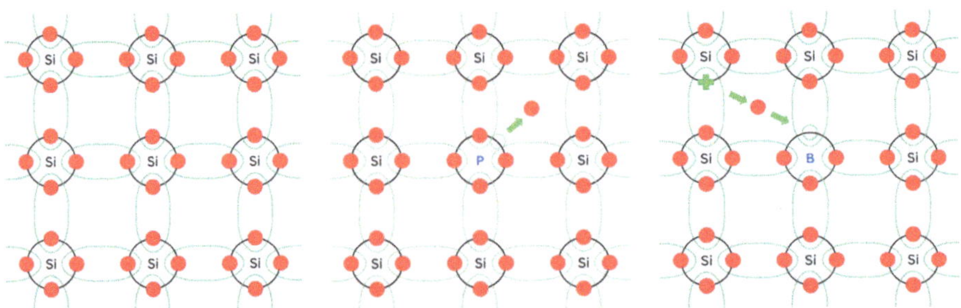


*a] Pure Silicon Crystal*
*A tidy lattice of silicon atoms where every electron is tied up in a bond. Only a handful break free at room temperature, so pure silicon conducts electricity only a little, just enough to call it a semiconductor rather than an insulator.*

*b] n-Type Silicon (Phosphorus-Doped)*
*Swapping in a single phosphorus atom gives the lattice one "spare" electron. That extra electron is free to roam, turning the silicon into an electron-rich (negative, n-type) material that conducts much more easily.*

*c] p-Type Silicon (Boron-Doped)*
*Replacing a silicon atom with boron leaves one bond short an electron, a "hole." Electrons from nearby atoms hop in to fill the gap, so the hole itself moves. This makes the lattice hole-rich (positive, p-type), providing an easy path for current in the opposite direction.*

**Fig. 9.1** N-type and P-type doping. (**a**) Pure Silicon Crystal: A tidy lattice of silicon atoms where every electron is tied up in a bond. Only a handful break free at room temperature, so pure silicon conducts electricity only a little, just enough to call it a semiconductor rather than an insulator. (**b**) n-Type Silicon (Phosphorus-Doped): Swapping in a single phosphorus atom gives the lattice one "spare" electron. That extra electron is free to roam, turning the silicon into an electron-rich (negative, n-type) material that conducts much more easily. (**c**) p-Type Silicon (Boron-Doped): Replacing a silicon atom with boron leaves one bond short an electron, a "hole." Electrons from nearby atoms hop in to fill the gap, so the hole itself moves. This makes the lattice hole-rich (positive, p-type), providing an easy path for current in the opposite direction

The beauty of doping is that we only need a very small amount to make a big difference, typically just a few dopant atoms for every million silicon atoms. It's like how a teaspoon of salt can transform the flavor of an entire pot of soup.

### 9.1.2 Creating Useful Electronic Structures

Doping isn't just about making silicon more conductive. Its real power comes from selectively doping different regions of the same silicon wafer differently. By creating adjacent regions of p-type and n-type silicon, we form what's called a **p-n junction,** which is the fundamental building block of semiconductor electronics.

At the boundary between p-type and n-type regions, something remarkable happens: electrons from the n-side diffuse into the p-side and holes from the p-side diffuse into the n-side, creating a **depletion region**, a narrow zone that resists current flow in one direction but allows it in the other. This is how we create diodes, which act like one-way valves for electricity.

By arranging multiple p-n junctions in specific patterns, we can create transistors that are the switches which form the foundation of all modern computing. For example, in a basic transistor, we might have an n-type region, followed by a thin p-type region, followed by another n-type region, which is an n-p-n transistor. By controlling the electrical conditions in the middle region, we can switch the transistor on or off to control the flow of current through the device.

Without the ability to create these differently doped regions, we couldn't build any of the billions of transistors that make up modern chips. Doping is truly what brings silicon to life electronically.

## 9.2 Diffusion Doping: Baking Impurities into Silicon

Now that we understand why doping is critical, let's explore how manufacturers actually get those dopant atoms into the silicon. The first method developed and still used in some processes today is **diffusion doping**.

### 9.2.1 The Basics of Diffusion

Diffusion works on a simple principle: at high temperatures, atoms naturally move from areas of high concentration to areas of low concentration. It's the same principle that causes a drop of food coloring to gradually spread through a glass of water, or the aroma of baking cookies to fill your entire house.

In semiconductor manufacturing, diffusion doping involves exposing silicon wafers to a dopant source at high temperatures (typically 900-1200 °C). The heat gives atoms enough energy to move, allowing dopant atoms to gradually penetrate the silicon crystal (Fig. 9.2).

### 9.2.2 The Diffusion Process in Action

Here's how a typical diffusion doping process works: Silicon wafers are loaded into a quartz tube or boat. The tube is placed in a special furnace and heated to around 1000 °C. A gas containing the dopant is introduced (e.g., phosphine for phosphorus doping or diborane for boron doping). The dopant atoms land on the silicon surface and begin to diffuse into the crystal. Over time (typically 30 min to several hours), the dopant penetrates deeper into the silicon. Finally, the wafers are slowly cooled and removed from the furnace.

It's very much like brewing tea, the longer you let the tea bag steep in hot water, the deeper the flavor penetrates. Similarly, the longer the diffusion process runs, the deeper the dopants penetrate into the silicon.

### 9.2.3 Controlling the Doping Profile

The concentration of dopants is highest at the surface and gradually decreases with depth, creating what engineers call a "diffusion profile." By controlling the temperature, time, and dopant gas concentration, manufacturers can adjust how deep the dopants go and how concentrated they are (Fig. 9.3).

For example, higher temperatures cause faster diffusion (dopants penetrate deeper in the same amount of time), while longer times at the same temperature also increase the penetration depth.

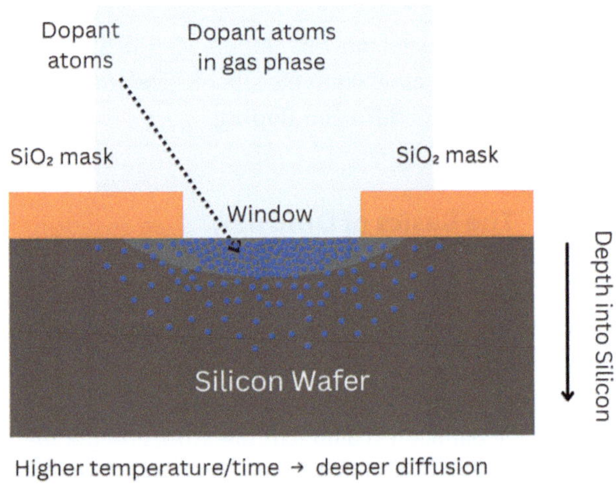


**Fig. 9.2** Diffusion doping process

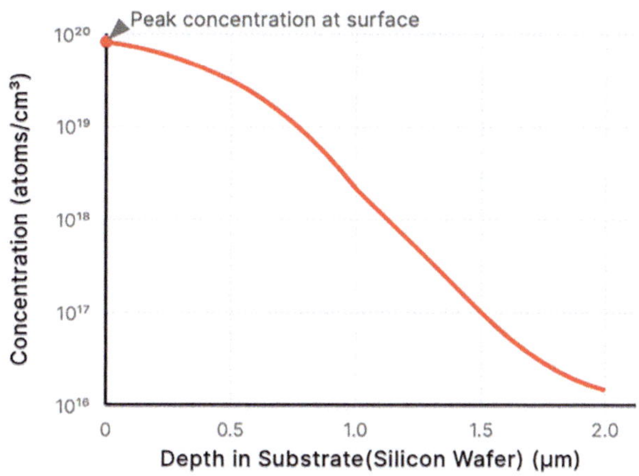


**Fig. 9.3** The concentration with diffusion doping is highest at the surface and decreases exponentially with depth

### 9.2.4 Selective Doping with Diffusion

Of course, in most cases, we don't want to dope the entire wafer uniformly. To create transistors and other devices, we need to dope specific regions while leaving others untouched. This is where diffusion doping works hand-in-hand with other processes we've already learned about.

First, an oxide layer is grown or deposited on the wafer (as we saw in Chap. 7). Using photolithography and etching (Chaps. 6 and 8), windows are opened in the oxide where doping is desired. During diffusion, the oxide acts as a mask, as dopants can only enter the silicon through the openings. After diffusion, the wafer has selectively doped regions exactly where needed.

This selective process was a key innovation in the "planar process" developed in the late 1950s, which made integrated circuits possible by allowing the creation of multiple transistors on a single piece of silicon.

### 9.2.5 The Historical Importance of Diffusion

In the early days of semiconductor manufacturing (1950s–1970s), diffusion was the primary method for doping silicon. Entire wafers would be processed in large diffusion furnaces, with carefully controlled atmospheres of dopant gases.

While diffusion is conceptually simple and requires relatively straightforward equipment, it does have limitations. The dopant profile is always a gradual curve, making it difficult to create very sharp transitions between doped and undoped regions. Also, as devices shrunk, the need for shallower, more precisely controlled doping profiles grew, which pushed the industry toward more advanced techniques.

## 9.3 Ion Implantation: Firing Dopants into the Wafer

As transistors grew smaller and manufacturing requirements became more stringent, a more precise method of doping emerged: **ion implantation**. If diffusion doping is like steeping tea, ion implantation is more like shooting tiny paintballs, providing much greater precision and control.

### 9.3.1 The Ion Implantation Process

Ion implantation uses a particle accelerator-like machine to ionize the dopant atoms (give them an electric charge), accelerate these ions to high speeds using electric fields, focus them into a beam, and "fire" this beam directly at the silicon wafer.

When these high-energy ions hit the silicon, they physically penetrate the crystal structure, embedding themselves at a certain depth depending on their energy. It's somewhat like firing small bullets into a target, the faster they're going, the deeper they'll penetrate (Fig. 9.4).

### 9.3.2 Precise Control of Doping

The beauty of ion implantation is the precise control it offers. By adjusting the acceleration energy, manufacturers can precisely control how deep the ions penetrate. Higher energy means deeper implantation. The number of ions fired (the "dose") determines the concentration of dopants. Want exactly $10^{15}$ dopant atoms per square centimeter? An ion implanter can deliver exactly that (Fig. 9.5).

**Fig. 9.4** Ion implantation

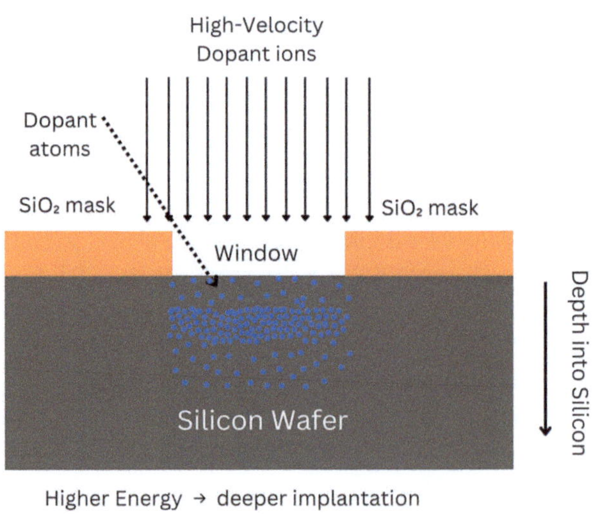

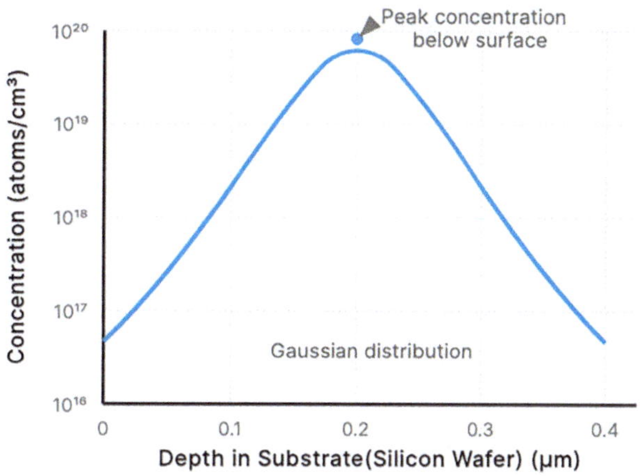


**Fig. 9.5** The concentration with ion implantation peaks at a characteristic depth below the surface, with a Gaussian distribution

Modern implanters scan the beam across the wafer or move the wafer through the beam to ensure extremely uniform doping across the entire surface. Unlike diffusion, which requires high temperatures during the doping process, ion implantation is performed at room temperature. This prevents existing dopants from moving around and keeps diffusion profiles sharp.

### 9.3.3 Selective Implantation

Like diffusion, ion implantation is typically done selectively. Areas that shouldn't receive dopants are protected using photoresist or other masking materials. Photoresist is applied to the wafer and patterned using photolithography. The resist acts as a shield, and ions cannot penetrate the thick resist layer. Only the exposed areas receive the implanted dopants. After implantation, the resist is removed.

The result is precisely doped regions exactly where needed, with sharp boundaries between doped and undoped areas.

### 9.3.4 The Machine Behind the Magic

An ion implanter is a sophisticated piece of equipment with several key components. The ion source generates ions from a gas or solid containing the desired dopant. A mass analyzer acts like a filter to ensure only the desired ion species reaches the wafer. The acceleration column speeds up the ions to the desired energy level. A beam scanner ensures the ions are distributed evenly across the wafer. Finally, the target chamber holds the wafer in position during implantation.

Modern ion implanters can accelerate ions to energies from a few kiloelectronvolts (keV) to several megaelectronvolts (MeV), allowing for a wide range of implantation depths.

### 9.3.5 A Technological Revolution

The introduction of ion implantation in the 1970s and its widespread adoption in the 1980s represented a significant technological leap for the semiconductor industry. It enabled more precise control over doping profiles, allowing for smaller transistors and more complex circuits.

By the 1980s, virtually all source and drain regions in logic transistors were formed using ion implantation, enabling feature sizes well below 1 μm, a critical step toward the billions of transistors we find on modern chips.

## 9.4 Activation and Thermal Annealing: Making Dopants Effective

Whether dopants are introduced via diffusion or ion implantation, there's often one more critical step needed: **thermal annealing**. This heating process serves two essential purposes: repairing crystal damage and activating dopants.

### 9.4.1 Repairing the Crystal Lattice

During ion implantation, high-energy ions slam into the silicon crystal, knocking silicon atoms out of position and creating damage. It is similar to firing small bullets into a neatly stacked wall of bricks, with some bricks getting displaced, leaving holes and creating a mess.

This damaged lattice is problematic because the dislocations and defects interfere with electrical properties, and many dopant atoms end up in "interstitial" positions (between regular lattice sites), where they don't contribute to the desired electrical behavior.

The solution is to heat the wafer, giving atoms enough energy to move around and rearrange themselves. This heating process is called annealing and it allows the silicon crystal to heal itself as atoms settle back into their proper positions (Fig. 9.6).

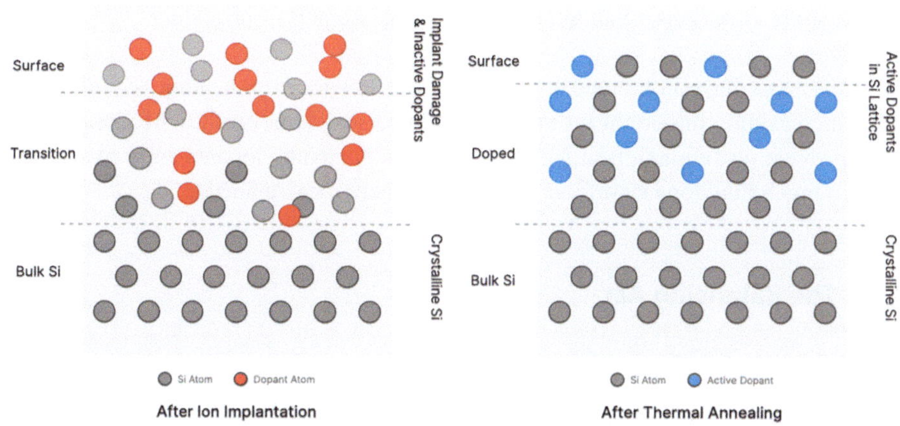


**Fig. 9.6** Crystal damage and annealing

### 9.4.2 Activating the Dopants

For dopants to effectively change silicon's electrical properties, they must be in the right position within the crystal lattice. Specifically, they need to replace silicon atoms in the lattice (called "substitutional" positions) rather than sitting between lattice sites.

Annealing provides the energy needed for dopants to move into these substitutional positions where they can donate or accept electrons as desired. Dopants that reach these positions are considered "electrically active" and they are actually doing their job of changing the semiconductor properties.

### 9.4.3 Annealing Techniques

As semiconductor manufacturing has evolved, so have annealing techniques. Traditional furnace annealing involves heating wafers in a furnace for minutes or hours at around 900–1000 °C. While effective at activating dopants and repairing damage, the long time at high temperature causes dopants to diffuse further, blurring the sharp profiles created by ion implantation.

**Rapid Thermal Annealing** (RTA) emerged as a solution, where wafers are quickly heated to high temperatures (typically 950–1050 °C) for just a few seconds using powerful lamps. This activates dopants while minimizing unwanted diffusion.

As technology advanced, even faster techniques were developed. **Spike annealing** is an even faster version of RTA, where the wafer temperature rapidly ramps up, reaches a peak temperature for just a split second, then quickly ramps down. **Flash annealing** uses an extremely brief flash of intense light to heat only the wafer surface for milliseconds, minimizing dopant diffusion. **Laser annealing** uses a laser to heat specific areas of the wafer for ultra-short durations (nanoseconds to microseconds), providing very localized annealing.

As transistors have continued to shrink, these advanced annealing techniques have become increasingly important for maintaining precise dopant profiles. Modern annealing aims to provide just enough heat for long enough to activate dopants and repair damage, but not so much that dopants diffuse away from where they're needed.

### 9.4.4 The Balancing Act

There's often a trade-off in annealing: Higher temperatures and longer times improve dopant activation and damage repair, but they also cause more dopant diffusion, blurring carefully placed dopant profiles.

Modern manufacturing processes often use multi-step annealing strategies, with different temperatures and durations optimized for specific layers and device requirements.

### 9.4.5 Bringing It All Together: The Doping Sequence

Let's see how doping typically fits into the semiconductor manufacturing process. A silicon wafer undergoes initial processing (cleaning, oxidation, etc.). Photolithography creates a pattern of exposed and protected areas. Ion implantation introduces dopants into the exposed regions. Thermal annealing repairs crystal damage and activates dopants. Further processing continues to create additional device structures.

This sequence is repeated multiple times throughout the manufacturing process, creating various doped regions for transistors and other devices. Each layer builds upon previous ones, gradually constructing the complete circuit with billions of components.

The precise control of where, how much, and how deep dopants go is what enables the incredible complexity of modern chips. Without these doping techniques, the digital revolution simply wouldn't have been possible.

With billions of transistors now patterned and doped on our wafer, we need to hook them up into circuits. In Chap. 10, we'll see how we create the intricate metal wiring (interconnect) that ties all these transistors together to form functional circuits.

**Summary**

- Doping is the process of adding precisely controlled amounts of impurity atoms to silicon, creating the electrically active regions that make transistors and other semiconductor devices possible.
- N-type doping (using elements like phosphorus or arsenic) adds extra electrons to the silicon, while p-type doping (using elements like boron) creates "holes" that act as positive charge carriers.

- Diffusion doping, the original method, uses high temperatures to drive dopant atoms into silicon from a gas source, like how tea diffuses into hot water.
- Ion implantation, the modern approach, accelerates dopant ions and fires them directly into the silicon, offering superior precision and control over dopant placement and concentration.
- After doping, thermal annealing repairs crystal damage from implantation and activates dopants by helping them move into the proper positions in the silicon lattice.

## Review Questions

1. In your own words, why is doping essential in turning a plain silicon wafer into a functioning electronic chip? What do p-type and n-type mean, and how do they relate to doping?
2. What are common dopant elements used for p-type and n-type silicon? How does each type of dopant change the charge carriers available in silicon?
3. Explain how photolithography helps ensure that only certain regions of the wafer get doped. What would happen if we doped the entire wafer uniformly with no patterns?
4. Think about control: which doping method (diffusion or implantation) gives finer control over how deep dopants go and how many are in the silicon? What trade-offs does the other method have (e.g., equipment complexity, uniformity)?
5. (Advanced thought) As transistors got smaller, junction depths had to get shallower. Which doping technique became crucial for making very shallow junctions, and why?

# Interconnect: Wiring Up the Chip

# 10

We've spent the last several chapters exploring how transistors are created on silicon wafers. But a single transistor or even billions of them cannot do anything useful unless they're connected together. That's where **interconnect** comes in, the intricate network of tiny metal wires that link transistors into functioning circuits and connect them to the outside world.

Think of the interconnect as the nervous system of the chip. If transistors are the neurons that process information, interconnects are the pathways that carry signals between them. Without this sophisticated wiring network, even the most advanced transistors would be useless, isolated components.

In this chapter, we'll explore how chip manufacturers create these complex wiring networks, building upward from the transistor layer in a multi-story architecture of metal and insulating materials.

## 10.1 From Transistor to First Metal: Contacts and the First Wiring Layer

Once all the transistors have been formed on the silicon wafer, we need to begin connecting them together according to the circuit design. But there is a problem: the transistors are covered by an insulating layer of silicon dioxide (essentially glass) that was used during the doping and gate formation processes. We need to create pathways through this insulating layer to reach the transistor terminals.


A. D. Watpade, *From Sand to Chip*, Synthesis Lectures on Engineering, Science, and Technology, https://doi.org/10.1007/978-3-032-18439-9_10

### 10.1.1 Making Contact

The first step in connecting transistors is to create what are called "**contacts**," which are small vertical connections that reach down through the insulating layer to touch specific parts of the transistors below, such as the source, drain, or gate.

To make contacts, we first apply photolithography to pattern small holes in the places where we need to connect to the transistors. Then we use etching (typically a highly selective plasma etch as described in Chap. 8) to carve through the insulating oxide layer, creating tiny contact holes that reach all the way down to the transistor terminals.

The contact etching process requires precise control to stop exactly when we reach the silicon or gate material without damaging the transistor. Once these holes are etched, we need to fill them with a conductive material to create the electrical connection.

Most modern processes first deposit a thin **barrier layer** (like titanium nitride (TiN)) to prevent unwanted reactions between the contact metal and silicon, followed by filling the holes with tungsten. Tungsten is commonly used for contacts because it provides excellent coverage in deep, narrow holes. Any excess tungsten on the wafer surface is then removed, leaving tungsten plugs only in the contact holes (Fig. 10.1).

### 10.1.2 The First Metal Layer

With the contacts in place, we're ready to create the first layer of actual wiring. This involves depositing a thin layer of metal across the entire wafer, then patterning it into the specific wire paths needed by the circuit design.

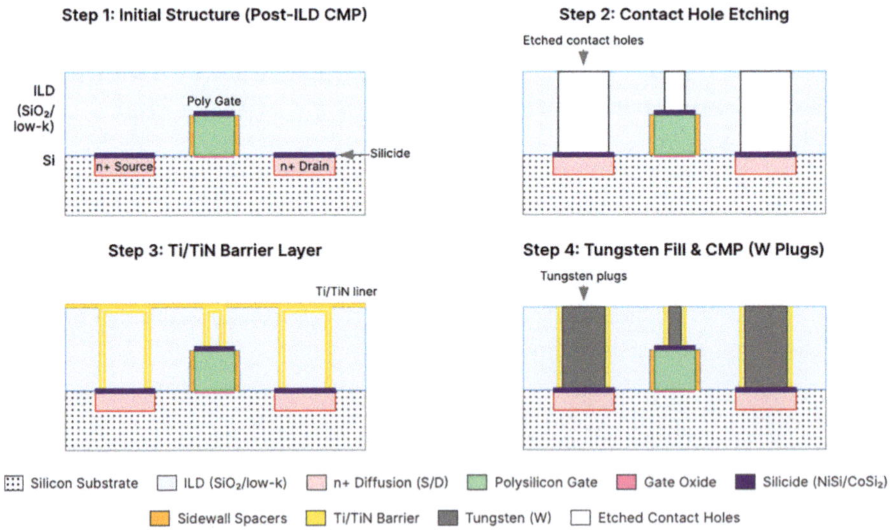


**Fig. 10.1** Contact formation

Traditionally, aluminum was the metal of choice because it's a good conductor, adheres well to silicon dioxide, and can be relatively easily deposited and patterned. The metal is typically deposited using a process called **sputtering,** which involves bombarding an aluminum target with high-energy ions that knock off atoms, which then settle onto the wafer in a thin, even layer.

Once the aluminum layer is deposited, we once again use photolithography to pattern the desired wire routes. The pattern is transferred to the aluminum using an etching process that removes the unwanted aluminum, leaving behind only the metal wires that follow our circuit design.

These wires now connect to the contact plugs we created earlier, establishing electrical pathways to the transistors below. Some wires might connect one transistor's output to another's input, while others might connect to power supply lines or eventually to the chip's external connections.

### 10.1.3 A Construction Analogy

If we think of building a chip like constructing a multi-story building, what we've done so far is:

1. Built the foundation and ground floor (the transistors)
2. Added support pillars (contacts) reaching up from the ground floor
3. Constructed the first-floor hallways and rooms (metal wires) that connect to those pillars

This first level of wiring allows for some connections between nearby transistors, but what about connections that need to travel longer distances or cross over other wires? For that, we'll need to build upward with additional floors of wiring.

### 10.1.4 Historical Perspective

It's worth noting that the earliest integrated circuits in the 1960s often had just a single layer of metal wiring. With only a few dozen transistors to connect, one layer was sufficient. As transistor counts grew from hundreds to thousands to millions and beyond, the wiring became increasingly complex and demanded multiple layers of interconnect.

## 10.2 Multilevel Metallization: Stacking Layers of Wiring

As chips grew more complex, a single layer of metal wiring quickly became insufficient. The problem is similar to traffic in a growing city when all roads are at ground level and intersections become inevitable. In electronic circuits intersecting wires would create unwanted electrical connections, i.e., short circuits.

### 10.2.1 The Need for Multiple Layers

Imagine trying to connect thousands of transistors using wires that all have to run in the same plane without crossing. It would be impossible without making the chip enormously large to accommodate all the detours needed. The solution, both in city planning and chip design, is to build upward, creating multiple levels of roadways or wiring that can pass over and under each other.

In modern chips, multiple layers of metal interconnect are stacked on top of each other, separated by insulating layers. This allows wires to cross over each other without making electrical contact. Lower metal layers generally handle local connections between nearby transistors, while upper layers handle longer-distance connections, power distribution, and clock signals that need to reach many parts of the chip (Fig. 10.2).

### 10.2.2 Building the Stack

Creating this multilevel wiring network follows a repeating process:

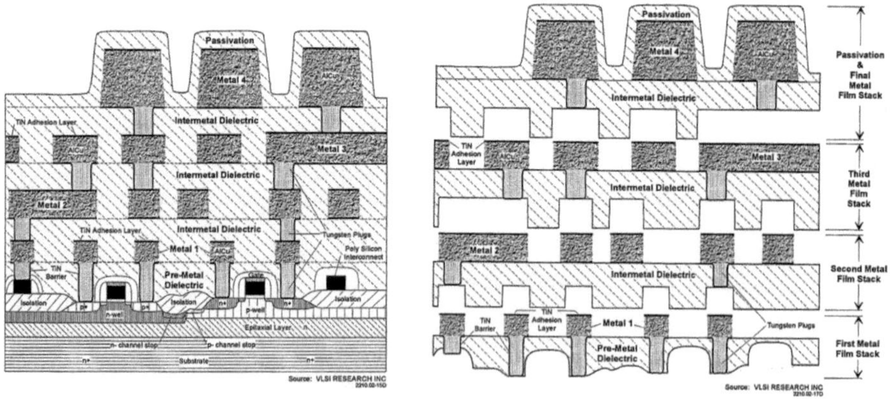


**Fig. 10.2** Cross section of multilevel metallization. Credit: VLSI

After the first metal layer is complete, we deposit another layer of insulating material (typically silicon dioxide or a similar dielectric) over the entire wafer. This insulates the first layer of metal from the next one we'll create.

Next, we need to create connections between metal layers. Similar to the contacts that connect to transistors, we create "**vias**," which are vertical connections between different metal layers. The process is similar; we use photolithography to pattern holes, etch through the insulating layer where we want connections, and fill the holes with metal.

With the vias in place, we deposit another layer of metal, pattern it into the desired wire paths, and the process continues. Each new metal layer can connect to layers below it through vias, allowing for complex three-dimensional routing of signals.

Modern processors might have 10–15 layers of metal, each with thousands or millions of wires, all precisely aligned and connected through vias to create the complete circuit. Lower layers typically have thinner, more densely packed wires for local connections, while upper layers have thicker wires for power distribution and global signals.

### 10.2.3 The Evolution from Aluminum to Copper

For decades, aluminum was the standard metal used for chip wiring. However, as wires became thinner and longer with increasing chip complexity, the electrical resistance of aluminum became a limiting factor. A higher resistance results in more signal delay, more power consumption, and more heat generation which are all undesirable in advanced chips.

In the late 1990s, IBM pioneered a significant shift by introducing copper interconnects for production chips. Copper offers about 40% lower electrical resistance than aluminum, allowing for faster signal transmission and lower power consumption.

However, switching to copper wasn't as simple as just substituting one metal for another. Copper has a tendency to diffuse into silicon and contaminate transistors, and it's also more difficult to pattern using traditional etching techniques. These challenges required developing entirely new processes for creating copper wiring, as we'll see in the next section.

## 10.3 Chemical-Mechanical Planarization (CMP): Keeping it Flat

As we build upward with multiple layers of interconnect, a critical challenge emerges, maintaining a flat surface. Each layer we add, metal wires, insulating materials, creates a more uneven, bumpy topography on the wafer. These bumps create serious problems:

- Photolithography requires a flat surface to maintain focus across the wafer
- Uneven surfaces can cause gaps or thin spots in deposited layers
- Wire thickness variations lead to inconsistent electrical properties
- Small bumps compound with each layer, eventually creating major topography issues

The solution to this problem came in the form of **Chemical-Mechanical Planarization (CMP)**, sometimes called "polishing", a process that literally smooths the wafer surface after certain steps.

### 10.3.1 How CMP Works

CMP is somewhat like sanding a rough wooden surface to make it smooth and even. The wafer is held face-down against a rotating polishing pad, with a slurry of abrasive particles and chemicals flowing between them. As the pad rotates, it gently wears down the high points on the wafer until the surface is completely flat.

The "chemical" part of CMP involves specialized slurry formulations that soften or react with specific materials, making them easier to remove. The "mechanical" part is the physical abrasion that wears away material. By controlling the chemistry, pressure, rotation speed, and timing, CMP can achieve incredible precision, removing just enough material to create a perfectly flat surface without damaging the structures below (Fig. 10.3).

### 10.3.2 The Damascene Process for Copper

CMP became especially important with the introduction of copper interconnects. Since copper is difficult to pattern using traditional etching techniques, manufacturers developed

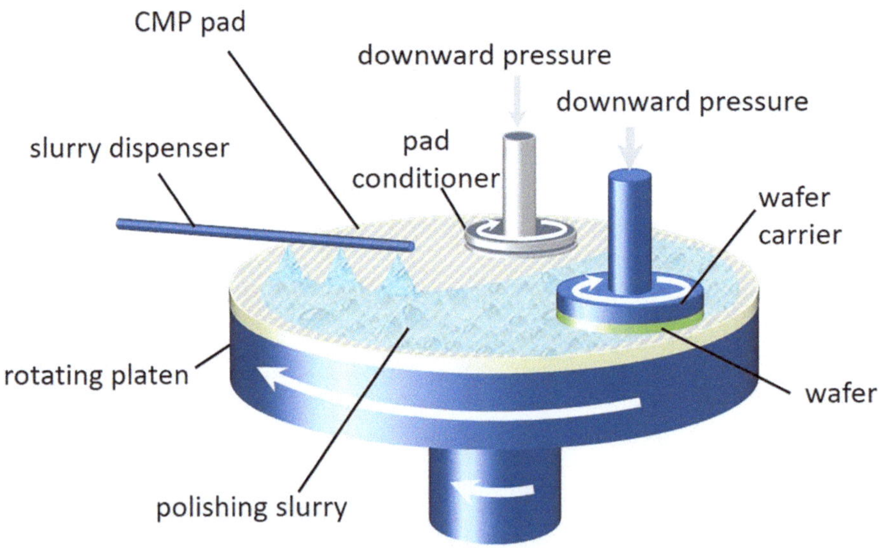


**Fig. 10.3** Chemical-Mechanical Planarization (CMP) process smoothing a wafer: a polishing pad flattens excess material to leave a level surface. Credit: Novacam

a new approach called the "**damascene process**" (named after the ancient Damascus metalworking technique of inlaying metal into grooves):

1. Deposit an insulating layer
2. Use photolithography and etching to create trenches (for wires) and holes (for vias) in the insulator
3. Deposit a thin barrier layer to prevent copper diffusion
4. Deposit copper, which fills the trenches and holes, but also covers the entire surface
5. Use CMP to polish away the excess copper, leaving copper only in the trenches and holes
6. The result is a flat surface with inlaid copper wires and vias

This "inlaid" approach is the opposite of traditional aluminum patterning. Instead of depositing metal and then etching away the unwanted portions, we create trenches where we want the metal and then fill them in. The CMP step is crucial because it removes all the excess copper, leaving a perfectly flat surface with metal wires embedded in the insulator.

For **dual damascene**, a more advanced version of this process, both the vias and the next layer's trenches are etched before the copper is deposited, allowing both to be filled simultaneously and then polished once.

### 10.3.3 A Revolutionary Technology

When CMP was first introduced, many engineers were skeptical about "sandpapering" delicate electronic components. But the process proved to be remarkably controllable and has become a cornerstone of modern chip manufacturing.

Without CMP, building the 10+ layers of interconnect in modern chips would be impossible. By the time you reach the upper layers, the accumulated topography would be too extreme for accurate lithography. CMP essentially gives manufacturers a fresh, flat surface to work with at critical steps in the process.

## 10.4 Interconnect Challenges and Innovations: Resistance, Capacitance, and Beyond

As chips have continued to advance, with smaller transistors and more layers of interconnect, the wiring itself has become a limiting factor in chip performance. Two key electrical properties create challenges as wires get thinner and more tightly packed:

### 10.4.1 Resistance and Capacitance Challenges

**Resistance**: As wires get thinner, their electrical resistance increases like a narrow pipe restricts water flow compared to a wide pipe. Higher resistance means signals travel more slowly and generate more heat.

**Capacitance**: When two conductors (like adjacent wires) are separated by an insulator, they form a capacitor that can store and release electrical charge. This capacitance between densely packed wires can cause signal delays and crosstalk (interference between signals).

Together, these effects create what engineers call "**RC delay**" (Resistance-Capacitance delay), which is a fundamental limit on how quickly signals can propagate through the chip's wiring. Even if transistors can switch faster, the interconnect delays can dominate overall chip performance.

### 10.4.2 Material Innovations

To address these challenges, the industry has introduced several key innovations:

**Copper Interconnect**: As we've seen, the switch from aluminum to copper reduced resistance by about 40%, significantly improving performance. This transition, pioneered by IBM in the late 1990s, has become universal in advanced chips.

**Low-k Dielectrics**: Traditional silicon dioxide has a relatively high dielectric constant (k-value), which increases capacitance between wires. Modern processes use low-k insulating materials that often contain carbon or even microscopic air pockets that reduce the dielectric constant and thus lower capacitance.

**Ultra Low-k and Air Gaps**: In the most advanced processes, engineers have even introduced "air gaps", which are actual tiny void spaces between wires. Since air has the lowest possible dielectric constant very close to 1, these air-insulated wires experience minimal capacitance.

### 10.4.3 Reliability Concerns

Beyond performance, reliability is a critical concern for interconnects. One particular challenge is **electromigration**, which is a phenomenon where the electrical current flowing through a wire gradually pushes metal atoms in the direction of the current flow. Over time, this can create voids (holes) in one part of the wire and hillocks (bumps) in another, eventually leading to open circuits or shorts.

Copper has better electromigration resistance than aluminum at the same current density, which is another reason for its adoption. Additionally, designers implement various safeguards against electromigration:

- Adding barrier layers around the copper

- Using wider wires for high-current paths
- Designing redundant vias for critical connections
- Limiting current densities in the layout design

### 10.4.4 Future Directions

As traditional interconnect scaling faces increasing challenges, researchers are exploring several alternatives for future chips:

- **Optical Interconnect**: Using light instead of electricity to transmit signals between different parts of a chip or between chips
- **3D Integration**: Stacking multiple chips and connecting them vertically with through-silicon vias (TSVs)
- **Advanced Packaging**: Moving some longer interconnections off the silicon die and onto packaging substrates with different interconnect technologies

While transistors often get the spotlight in discussions of chip technology, interconnect innovations have been equally crucial to the continued advancement of semiconductor performance. The ability to efficiently connect billions of transistors with minimal delay is what truly enables modern computing.

Now that our chips' transistors are wired together into functional circuits, we face one more challenge: getting those circuits out to the real world. A finished wafer contains many tiny chips that must be separated, and each chip must be packaged so it can be handled and connected to other devices. In Chap. 11, we'll see how chips are packaged. It is the final step that turns a fragile silicon die into a usable electronic component.

**Summary**

- An interconnect is the network of metal wiring that connects transistors together to form functional circuits, built as a multilayer structure above the silicon transistors.
- The interconnect process begins with creating contacts to the transistors, followed by patterning the first metal layer into wires that connect various transistor terminals according to the circuit design.
- Modern chips use multiple layers of metal interconnect (often 10–15 layers), connected by vertical vias, to create complex three-dimensional wiring networks that allow for billions of connections without wires shorting.
- The industry transition from aluminum to copper interconnects in the late 1990s reduced electrical resistance, while chemical-mechanical planarization (CMP) enabled the damascene process for patterning copper and maintaining flat surfaces between layers.

- As interconnects have scaled down, challenges with resistance, capacitance, and reliability have driven innovations in materials (like low-k dielectrics) and design techniques to ensure signals can still travel efficiently through increasingly complex wiring networks.

## Review Questions

1. What is the role of interconnect in an integrated circuit, and how does it relate to the transistors on the chip?
2. Describe the process of making a first-layer metal interconnect on a chip. How do we create contacts to the silicon and then form the first metal wiring?
3. Why do modern chips require multiple layers of metal wires? What would happen if we tried to route all connections in just one layer?
4. Imagine you have a bumpy surface and you need to add another layer of circuit on top. What problems might the bumps cause in manufacturing and circuit performance? How does CMP address these problems?
5. What are "low-k" dielectrics, and why were they introduced in interconnect structures?
6. Define electromigration in simple terms. Why is it a concern for tiny metal wires in chips, and how has the shift to copper or other measures improved interconnect reliability?

# Chip Packaging: Encasing and Connecting the Chip to the World

# 11

We've followed the journey of how silicon wafers are processed to create transistors, interconnected with intricate metal wiring to form functional circuits. But at this point, what we have is still a bare silicon wafer with dozens or hundreds of identical chips (also called **dies**) built on its surface. These chips are incredibly small, fragile, and have microscopic connection points that can't interface directly with electronic devices.

**Packaging** is the crucial step that transforms these vulnerable silicon chips into the robust, usable components we find in electronic devices. Packaging serves multiple essential purposes:

- Protects the delicate silicon die from physical damage, moisture, and contamination
- Provides a way to connect the microscopic circuits on the chip to the macroscopic world of circuit boards
- Helps dissipate heat generated by the chip during operation
- Makes the chip easy to handle, test, and assemble into electronic products

In this chapter, we'll explore the fascinating process of how those silicon dies become the packaged chips that power our computers, smartphones, and countless other devices.

## 11.1 Wafer Dicing: From Wafer to Individual Dies

After all the fabrication processes are complete, we have a wafer full of identical chip patterns, though some might have defects. The first step in packaging is to separate these chips from one another, a process called **wafer dicing**.



A. D. Watpade, *From Sand to Chip*, Synthesis Lectures on Engineering, Science, and Technology, https://doi.org/10.1007/978-3-032-18439-9_11

### 11.1.1 The Dicing Process

Imagine a chocolate bar with pre-scored sections. You snap along those lines to get individual pieces. Wafer dicing follows a similar principle, but with much greater precision.

During the chip design phase, engineers include narrow spaces between each chip design called "**scribe lines**" or "**streets**." These streets contain no critical circuits as they are essentially sacrificial spaces where the wafer will be cut. The streets typically contain some test structures and alignment marks but nothing that needs to survive the cutting process (Fig. 11.1).

To prepare for dicing, the wafer is first attached to a sticky film stretched over a metal frame, similar to how an embroidery hoop holds fabric taut. This adhesive film serves an important purpose: it holds all the individual chips in their original positions after they're cut, making them easier to handle and retrieve.

The actual cutting is typically done in one of two ways:

- **Diamond Saw Dicing**: A precision diamond-tipped circular saw, spinning at high speed, cuts along the streets between the chips. Water is sprayed during cutting to cool the blade and wash away silicon dust. The saw is extremely thin (around 50 μm or less) and cuts completely through the wafer, but not through the adhesive film underneath.
- **Laser Dicing**: For more advanced or delicate chips, lasers may be used instead of mechanical saws. The laser beam ablates (vaporizes) the silicon along the scribe lines. Laser dicing can create narrower streets and cause less mechanical stress to the chips.

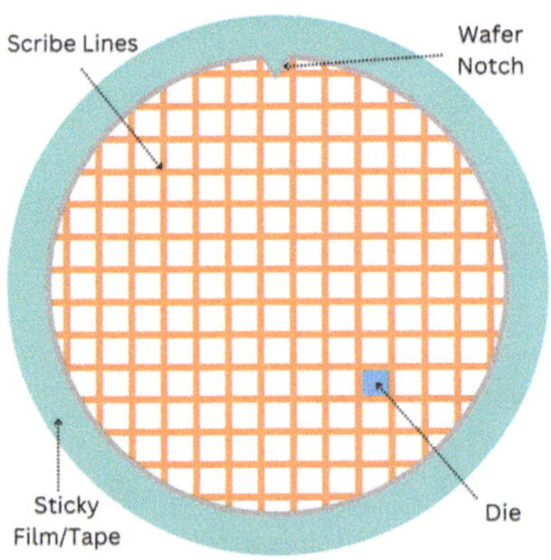


*Schematic Representation of Wafer Layout (Scribe Lines and Notch Enlarged for Visibility)*

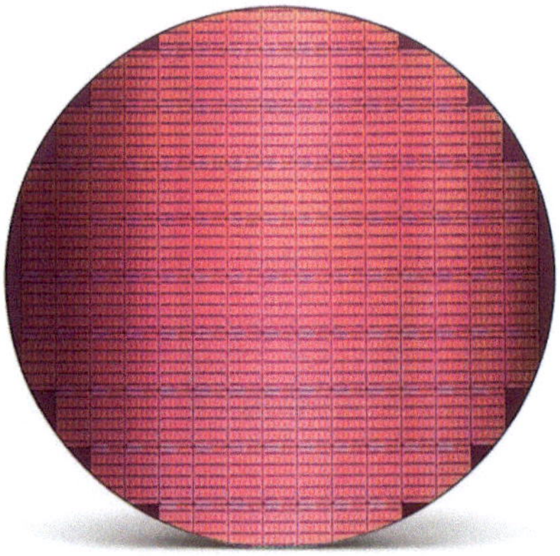

*Actual Silicon Wafer Showing Dies and Scribe Lines (Credit: Intel Corporation)*

**Fig. 11.1** Wafer with scribe lines. Credit: Intel Corporation

After dicing, what we have is the original wafer now cut into individual dies, still held in their original positions by the adhesive film. The next step is to pick up each good die and transfer it to its package. This is done with specialized equipment that can pick up the tiny dies with great precision, often using vacuum tools that work like miniature suction cups.

### 11.1.2 Wafer Maps and Die Selection

It's worth noting that before dicing, the wafer usually undergoes electrical testing (called wafer probe or wafer sort). This testing identifies which dies are good and which have defects. The results create a "wafer map" showing the location of functional and non-functional dies.

After dicing, only the good dies that are identified by the wafer map are selected for packaging. This saves the cost of packaging defective chips that would fail later tests anyway. However, the details of testing belong in another chapter, as our focus here is on the packaging process.

## 11.2 Traditional Packaging: Wire Bonding and Leadframe Packages

Once we have individual silicon dies, we need to connect them to a package that protects the die and provides electrical connections to the outside world. For decades, the most common approach has been wire bonding with a leadframe package.

### 11.2.1 The Leadframe

The leadframe is a thin, flat piece of metal (typically copper or a copper alloy) that's been stamped or etched into a specific pattern (Fig. 11.2). It serves as the skeleton of the package and has several important parts:

- A center pad (die paddle) where the silicon die will be mounted
- Leads or pins that extend from the frame to the outside of the package
- A supporting frame that holds everything together during assembly

### 11.2.2 Die Attach

The first step is to secure the silicon die to the center pad of the leadframe. This is called die attach or die bonding. A small amount of adhesive (usually a special conductive or non-conductive epoxy) is placed on the die paddle, and the die is precisely positioned and pressed into place. The adhesive is then cured, often by heating, to create a strong bond.

**Fig. 11.2** Typical leadframe

For power devices that generate significant heat, a thermally conductive material like solder might be used instead of epoxy to help transfer heat from the die to the leadframe and eventually to the outside world.

### 11.2.3 Wire Bonding

Now comes the critical step of connecting the die to the leadframe electrically, and this is where wire bonding comes in. Around the perimeter of the silicon die are small metal pads, **bond pads**, that connect to the circuits inside. These pads need to be connected to the inner ends of the leadframe's pins.

Wire bonding creates these connections using extremely fine wires, typically made of gold, aluminum, or copper. These wires are thinner than a human hair and are usually between 15 and 33 μm in diameter (Fig. 11.3).

The wire bonding process uses specialized machines that perform a sophisticated "stitching" operation:

1. The wire bonding tool (called a capillary) positions itself over a bond pad on the chip
2. It presses down with the wire, applying ultrasonic energy and/or heat to create a bond between the wire and the pad
3. The tool moves to the corresponding lead on the leadframe, trailing the wire
4. It creates a second bond, connecting the wire to the lead
5. The wire is cut, and the process repeats for each connection

Modern wire bonders can make thousands of these connections per hour with remarkable precision. Each bond must be reliable. A single broken wire can render the entire chip useless.

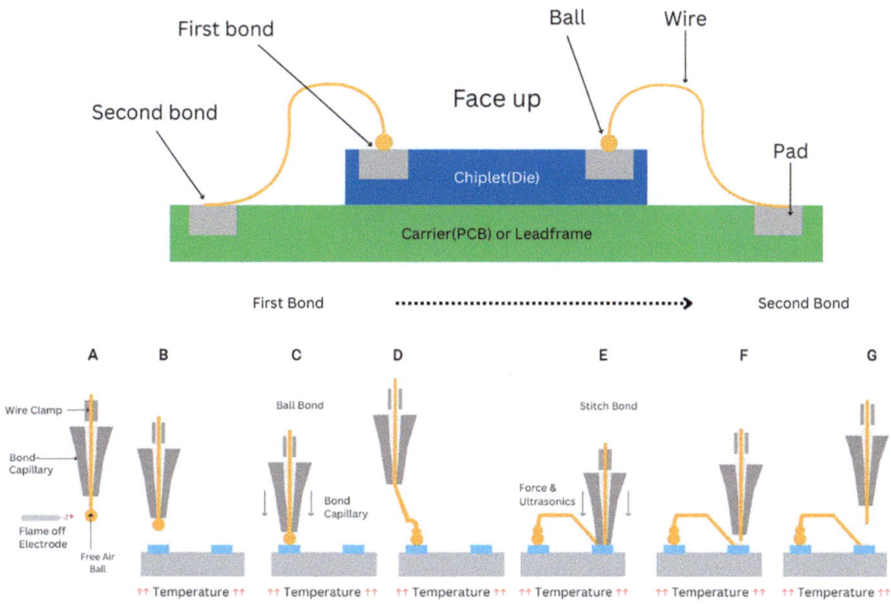


**Fig. 11.3** Wire bonding process. Credit: Protoexpress

### 11.2.4 Encapsulation

With the die attached and all wire bonds in place, the next step is to protect these delicate components from the environment. This is done through **encapsulation,** which involves enclosing the die, wires, and inner portions of the leadframe in a protective material.

For most packages, this involves transfer molding, where the assembled leadframe is placed in a mold, and an epoxy compound is injected under heat and pressure. This compound flows around the die and wire bonds, then hardens to form a solid body. Only the outer portions of the leads remain exposed to provide connections to the circuit board.

For some applications, ceramic packages or metal cans might be used instead of plastic molding, especially for chips that need better heat dissipation or hermetic sealing (complete protection from moisture).

### 11.2.5 Trim and Form

After molding, the leadframe is still connected to its outer frame, and the leads are still attached to the supporting structure. In the trim and form step, these excess parts are cut away, and the leads are bent into their final shape.

Different package types have different lead configurations:

- **Dual In-line Package (DIP)**: The classic "chip" most people picture, with two rows of pins designed to insert through holes in a circuit board.

- **Quad Flat Package (QFP)**: A thin, square package with leads extending from all four sides, bent downward in a "gull-wing" shape for surface mounting.
- **Small Outline Integrated Circuit (SOIC)**: A smaller version of the DIP with leads bent outward for surface mounting.

Each of these traditional package types relies on the same basic concept: a leadframe with wire bonds connecting to a silicon die, encapsulated in a protective material.

### 11.2.6 Historical Perspective

Traditional wire-bonded packages have been the industry standard for decades. If you opened up electronics from the 1970s through the 1990s, nearly every chip would use this packaging approach. Even today, many integrated circuits still use wire bonding because it's cost-effective and well-suited for many applications.

However, as chips became more complex with higher performance requirements, the limitations of wire bonding became apparent. The long, thin wires introduce inductance and resistance that can slow down signals. Additionally, wire bonding generally limits connections to the perimeter of the chip, constraining the total number of connections possible.

These limitations led to the development of more advanced packaging methods, as we'll see in the next section.

## 11.3 Flip-Chip and Ball Grid Array: Connecting with Solder Bumps

As integrated circuits became more complex with more input/output connections needed, traditional wire bonding faced limitations. Wire bonds can only be placed around the perimeter of the chip, limiting the total number of connections. Additionally, the long wires introduce electrical parasitics that can slow down high-speed signals.

Flip-chip technology emerged as a solution to these challenges, fundamentally changing how chips connect to their packages.

### 11.3.1 The Flip-Chip Concept

In **flip-chip** packaging, instead of using wire bonds, the chip is literally flipped upside down and directly connected to the package substrate (Figs. 11.4 and 11.5). Small solder bumps or copper pillars are formed on the bond pads of the chip, and these bumps make direct contact with corresponding pads on the substrate when the chip is flipped over and attached.

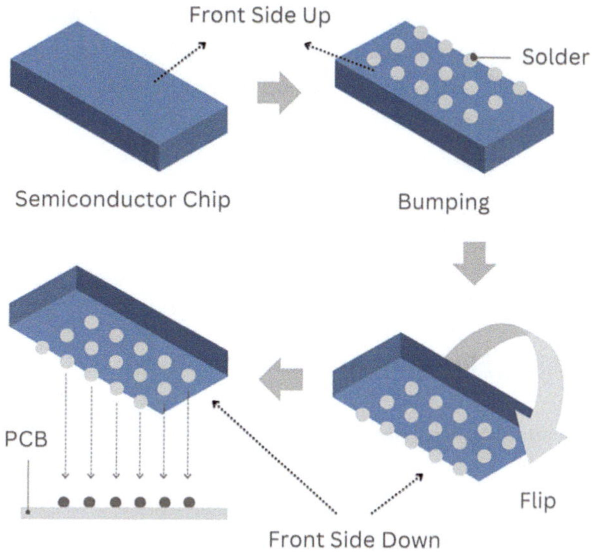


**Fig. 11.4** Flip-chip bonding

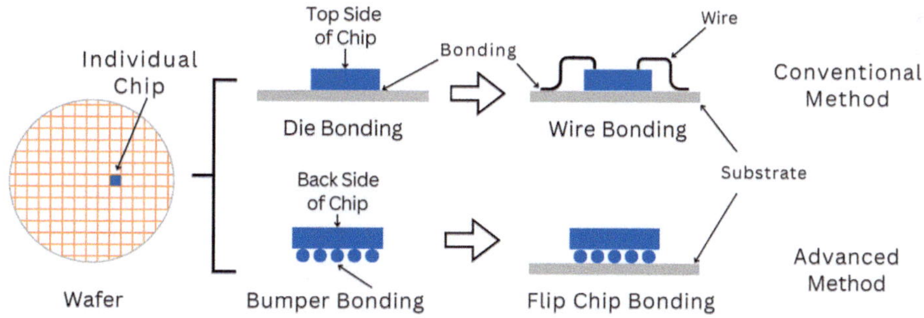


**Fig. 11.5** Wire-bonding vs flip-chip bonding

This approach offers several significant advantages:

- Connections can be made across the entire surface of the chip, not just around the edges
- The connection paths are much shorter than wire bonds, improving electrical performance
- Many more connections are possible in the same chip area
- Heat can transfer more directly from the chip to the package

### 11.3.2 Creating the Bumps

The solder bumps for flip-chip packaging are typically formed while the chips are still on the wafer, before dicing. This process, called **wafer bumping**, involves several steps:

1. A special metal layer called **Under Bump Metallization (UBM)** is deposited on the bond pads to create a surface suitable for attaching solder
2. Solder is deposited on each pad, either through evaporation, electroplating, or by placing pre-formed solder balls
3. The solder is melted briefly (reflowed) to form well-shaped bumps

After bumping, the wafer is diced as described earlier, resulting in individual dies with solder bumps ready for attachment.

### 11.3.3 Flip-Chip Attachment

To attach a flip-chip to its package, the process works as follows:

1. The bumped die is picked up and flipped over (circuit side down)
2. It is precisely aligned with the corresponding pads on the package substrate
3. The assembly is heated to melt the solder bumps, creating electrical and mechanical connections
4. An **underfill** material (typically an epoxy) is injected to fill the space between the die and substrate
5. This underfill is cured to provide mechanical strength and protection

The underfill is crucial because it protects the solder connections and helps distribute stress caused by thermal expansion differences between the silicon die and the substrate.

### 11.3.4 Ball Grid Array (BGA) Packages

Flip-chip technology is commonly used in conjunction with **Ball Grid Array (BGA)** packages. A BGA package has a grid of solder balls on its bottom surface instead of leads around the edges. These balls serve as the connections to the circuit board (Fig. 11.6).

Inside a typical flip-chip BGA package:

- The flip-chip die is attached to a package substrate (a small, multilayer circuit board).
- The substrate redistributes the connections from the fine-pitch bumps on the chip to the coarser grid of solder balls underneath.
- Often, a metal lid or heat spreader is attached to the back of the die to help dissipate heat.

BGAs have become the dominant package type for high-performance chips like CPUs, GPUs, and system-on-chips (SoCs) in smartphones. They allow for hundreds or even thousands of connections in a relatively small area.

*Top view of a flip-chip BGA package showing the metal heat spreader that helps cool the chip*

*Bottom view of a flip-chip BGA package showing the grid of solder balls that connect to the circuit board. (Credit: Intel Corporation)*

**Fig. 11.6** Ball grid array package. Credit: Intel Corporation

### 11.3.5 Industry Adoption

IBM pioneered flip-chip technology for its mainframe computers in the 1960s, but it wasn't until the 1990s that the technology became widespread in consumer electronics. Today, virtually all high-performance processors use flip-chip packaging.

The flip-chip approach has continued to evolve, with copper pillars often replacing solder bumps for finer pitch connections, and advanced substrate technologies enabling even greater connection densities.

## 11.4 Advanced and 3D Packaging: Chiplets and Stacked Dies

While traditional packaging methods treat each silicon die as a separate entity, advanced packaging technologies are blurring the line between chip and package by integrating multiple dies into unified systems. This approach is increasingly important as the semiconductor industry seeks ways to continue advancing performance beyond the limitations of conventional silicon scaling.

### 11.4.1 Multi-Chip Modules and System-in-Package

One of the simplest forms of advanced packaging is the **multi-chip module** (**MCM**), where multiple dies are mounted side by side on a common substrate and packaged together. This evolved into what is now often called **System-in-Package** (**SiP**), which

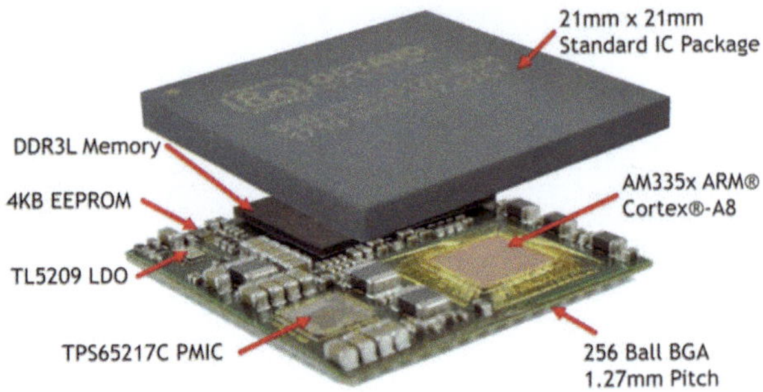


**Fig. 11.7** Example of system-in-package. Credit: Octavo Systems

might combine different types of chips like processors, memory, sensors, and communication chips into a single package.

In a smartphone, for example, you might find a SiP that contains the application processor, memory chips, and power management circuits all in one package instead of being separate components on the circuit board (Fig. 11.7).

### 11.4.2 The Chiplet Revolution

Traditionally, chip designers have focused on creating monolithic dies, single pieces of silicon containing all the necessary functions. However, as chips have grown larger and more complex, manufacturing them as a single die has become challenging:

- Larger dies have lower manufacturing yield (more chance of defects)
- Different functions might benefit from different manufacturing processes
- Integrating everything on one die is increasingly expensive

The **chiplet** approach addresses these challenges by breaking down what would have been one large chip into multiple smaller chips (chiplets), each optimized for its specific function, and then packaging them together to work as a unified system.

For example, AMD's Ryzen processors use multiple core chiplets (containing the CPU cores) combined with a separate I/O die that handles memory and peripheral connections. This approach allows AMD to optimize production, as the core chiplets use the most advanced process technology for performance, while the I/O die uses a more mature process better suited for its functions.

### 11.4.3 2.5D Integration with Silicon Interposers

To connect chiplets with the high density of connections they require, a technology called **2.5D integration** emerged. This approach uses a **silicon interposer,** essentially a piece of silicon with microscopic wiring but no active transistors, as a high-density connection platform.

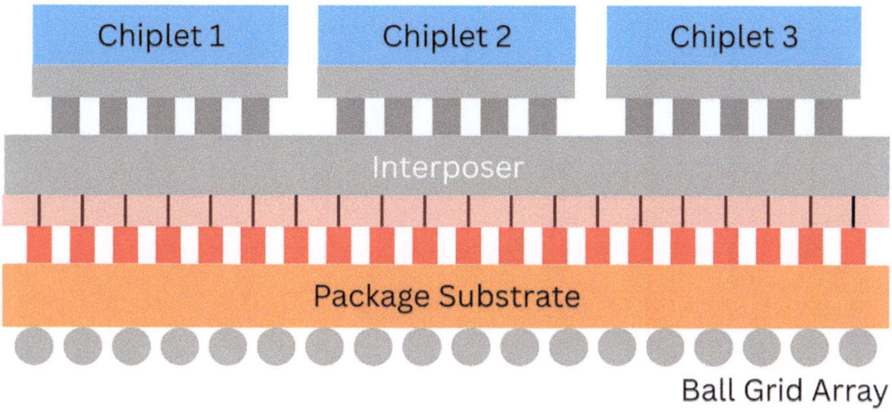


**Fig. 11.8** 2.5D integration with silicon interposer

Chiplets are mounted on this interposer using flip-chip technology, and the interposer provides the fine-pitch wiring between them. The interposer itself is then mounted on a package substrate.

A prime example of this technology is high-bandwidth memory (HBM) used with graphics processors. HBM stacks memory dies vertically (which we'll discuss next), and these memory stacks are placed alongside a GPU die on a silicon interposer, allowing thousands of connections between them (Fig. 11.8).

### 11.4.4 3D Stacking and Through-Silicon Vias

The most advanced form of packaging literally adds a third dimension by stacking dies directly on top of each other. This approach, called **3D stacking**, relies on a technology known as **through-silicon vias (TSVs)**.

TSVs are vertical connections that pass completely through a silicon die, allowing signals to travel directly from one die to another in the stack. Creating TSVs involves etching small holes completely through the silicon, then filling them with metal to form the connections (Fig. 11.9).

The advantages of 3D stacking are significant:

- Dramatically reduced connection distances between chips
- Much higher bandwidth between stacked components
- Smaller overall footprint for the same functionality
- Potential for mixing different types of chips in the same stack

High-bandwidth memory (HBM) is a prominent example of 3D stacking, with multiple DRAM dies stacked on top of each other and connected by TSVs. Intel's Foveros technology is another example, allowing logic chips to be stacked on top of each other.

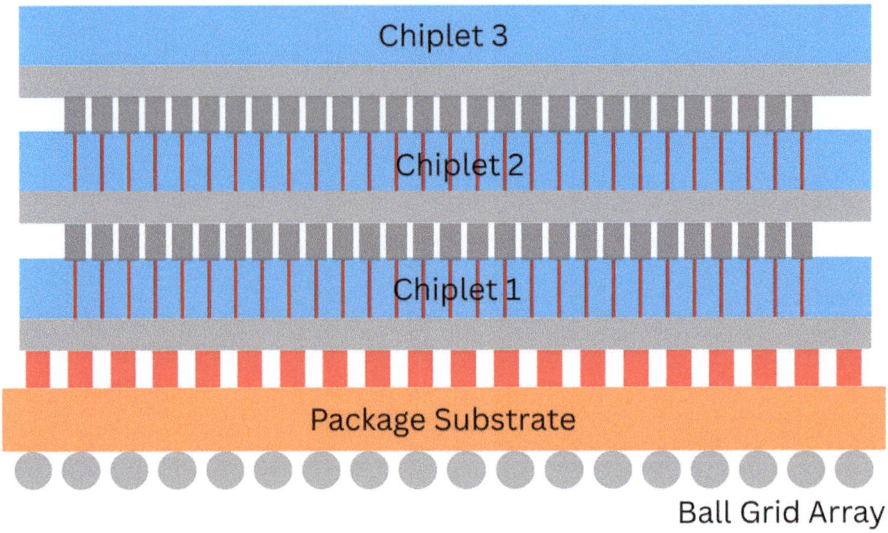


**Fig. 11.9** 3D stacking with through-silicon vias

### 11.4.5 Package-on-Package

A simpler form of 3D packaging that doesn't require TSVs is **Package-on-Package (PoP)**, where complete packages are stacked on top of each other. This is commonly used in smartphones, where the memory package is stacked on top of the processor package, saving valuable space on the circuit board.

The bottom package typically has a processor or system-on-chip, while the top package contains memory. Connections between packages are made through an array of solder balls around the periphery.

### 11.4.6 Packaging as the New Frontier

Advanced packaging has become a crucial frontier in semiconductor innovation. As traditional transistor scaling (making transistors smaller) becomes more challenging and expensive, creative packaging solutions offer an alternative path to improving performance, reducing power consumption, and adding functionality.

Industry leaders now talk about "More than Moore" as they explore ways to advance semiconductor technology beyond the traditional scaling approach defined by Moore's Law. Advanced packaging is at the heart of this new paradigm, as it enables heterogeneous integration of specialized chips rather than attempting to put everything on a single piece of silicon.

From Apple's M-series Ultra chips that fuse two dies together using advanced packaging, to AMD's chiplet-based approach for high-performance computing, to memory stacking for greater bandwidth, advanced packaging is transforming how we design and build semiconductor products.

### 11.4.7 Completing the Manufacturing Journey

At this point, we've transformed common sand into a packaged integrated circuit ready to be mounted on a circuit board. This journey includes purification, crystal growth, fabrication of transistors and wiring, and finally packaging. Each step required incredible precision, specialized equipment, and sophisticated techniques to turn one of Earth's most common materials into some of the most complex devices humans have ever created.

While we've completed our exploration of the manufacturing process itself, two critical questions remain. How do manufacturers ensure that this complex process yields working chips reliably? And what challenges and innovations lie ahead as we push semiconductor technology beyond its current limits? We'll address these topics in the final chapters of our journey.

In the next chapter, we'll explore how semiconductor manufacturers ensure that their complex processes consistently produce working chips. We'll learn about yield management, defect detection, and process control, the crucial disciplines that transform semiconductor manufacturing from an art into a reliable science that can produce billions of functional chips every year.

**Summary**

- Chip packaging transforms fragile silicon dies into robust components by protecting the die from damage, providing electrical connections to the outside world, and helping with heat dissipation.
- Wafer dicing separates individual chip dies from the processed wafer using diamond saws or lasers that cut along predefined streets between the chips.
- Traditional packaging uses wire bonding to connect the silicon die to a leadframe with fine gold, aluminum or copper wires, then encapsulates the assembly in plastic or ceramic.
- Flip-chip packaging improves on wire bonding by flipping the chip upside down and connecting it directly to the package substrate using an array of solder bumps, enabling more connections and better performance.
- Advanced packaging technologies like chiplets, 2.5D integration with interposers, and 3D stacking with through-silicon vias allow multiple dies to work together as a unified system, providing a new dimension for semiconductor advancement.

**Review Questions**

1. What are the main purposes of packaging a semiconductor chip? List at least two reasons why a bare silicon die is not used on a circuit board without a package.
2. Describe how a wafer is diced. Why are there usually "empty" spaces or scribe lines between individual chip circuits on a wafer?

3. What does BGA stand for, and why are BGAs commonly used for advanced chips?
4. What is meant by 2.5D and 3D packaging? Give one example of each.
5. Why might a company choose to use multiple smaller chiplets in a package instead of one large monolithic chip? What are some advantages of this chiplet approach?

# Yield, Defects, and Process Control: Making It Work, Every Time 12

Throughout the previous chapters, we've explored the intricate processes involved in creating semiconductor chips, from purifying silicon to fabricating transistors to packaging the final product. But there's a crucial factor we haven't yet discussed that determines whether all this effort results in profitable products, **yield**.

The semiconductor industry invests billions of dollars in fabrication facilities (fabs) filled with incredibly precise equipment. Yet, for all this technology, not every chip that comes off the production line will work correctly. Understanding why chips fail, how to detect problems early, and strategies to maximize the number of working chips is what this chapter is all about.

We'll explore the concept of yield and why it's so critical to semiconductor manufacturing economics. We'll look at the various defects that can plague production, the sophisticated statistical methods used to monitor processes, and some fascinating real-world examples of how manufacturers have solved yield challenges.

## 12.1 What Is Yield?

At its most basic, yield is simply the percentage of good chips produced from a wafer or batch of wafers. If we process a wafer with 500 individual chips (dies) on it, but only 400 of them pass all tests, the yield is 80% (Fig. 12.1).

### 12.1.1 The Economics of Yield

To understand why yield is so critically important, let's use a simple baking analogy. Imagine you're making cookies for a bake sale. Each batch requires the same ingredients,



A. D. Watpade, *From Sand to Chip*, Synthesis Lectures on Engineering, Science, and Technology, https://doi.org/10.1007/978-3-032-18439-9_12

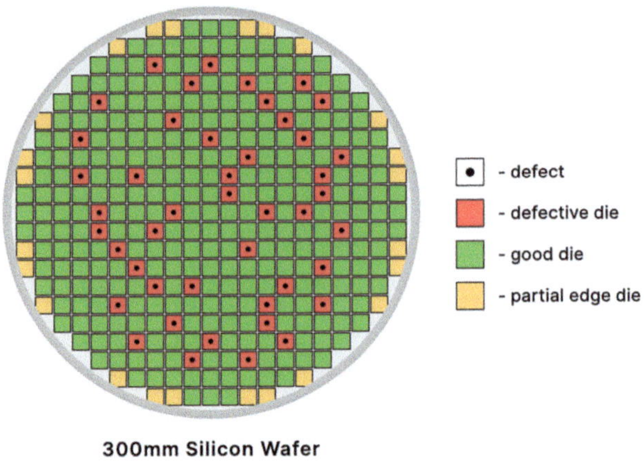


**Fig. 12.1** Wafer map showing die yield: good dies (green) and defective dies (red) across a wafer after testing

**Fig. 12.2** Yield impact on chip cost

preparation time, and oven energy, whether you successfully make ten cookies or just eight because two get burned.

If your ingredients and time cost $10 for a batch, each cookie from a perfect batch costs $1. But if two cookies are always ruined, your effective cost rises to $1.25 per good cookie, a 25% increase! The same math applies to semiconductor manufacturing but with much higher stakes.

A single 300 mm wafer can cost thousands of dollars to process through a modern fab. If we assume a wafer costs $5000 to manufacture and contains 500 potential chips, then with a perfect 100% yield (which never happens in reality), each chip would cost $10. With an 80% yield, that same chip now costs $12.50. If yield drops to 50%, the cost jumps to $20 per chip (Fig. 12.2).

In the real world, semiconductor manufacturing is fiercely competitive, and profit margins can be thin. A company with 70% yield might struggle to compete with a rival achieving 85% yield, as their manufacturing costs would be significantly higher for essentially the same product.

### 12.1.2 Die Yield vs. Wafer Yield

When discussing semiconductor yield, we typically distinguish between:

- **Die Yield**: The percentage of good individual chips (dies) on a wafer. This is what we calculated in our 400/500 = 80% example above.
- **Wafer Yield**: The percentage of wafers that successfully complete the entire manufacturing process. Some wafers might be scrapped partway through if major problems are detected.

Both types of yield matter. A fab might process thousands of wafers per month, so even losing a small percentage to wafer yield issues can be costly. But die yield usually has the larger economic impact, as it directly affects how many saleable products result from each expensive wafer.

### 12.1.3 Historical Perspective

Early in the semiconductor industry, yields were remarkably low. When Intel made its first microprocessors in the early 1970s, yields were often in the single digits, which meant more than 90% of chips didn't work. Today, mature processes routinely achieve yields above 90%.

This dramatic improvement didn't happen by accident. It came through decades of increasingly sophisticated methods to control the manufacturing environment, detect problems early, and design chips with yield in mind.

### 12.1.4 The Yield Improvement Journey

As a manufacturing process matures, its yield typically follows a predictable path:

1. **Development Phase**: When a new process is first created, yields are very low as engineers work out the basic kinks
2. **Ramp Phase**: As the process moves into production, yields climb rapidly as major issues are solved
3. **Mature Phase**: Yield improvements slow but continue through incremental refinements

4. **End-of-Life Phase**: Eventually, the process reaches its practical yield ceiling

For leading-edge processes pushing the boundaries of what's physically possible, this journey is particularly challenging. A new process node (like moving from 7 nm to 5 nm technology) often starts with significantly lower yields than the mature process it's replacing, then gradually improves.

Given the enormous economic impact of yield, even small improvements can justify major investments. A 1% yield improvement on a high-volume product can be worth millions of dollars annually.

## 12.2 Sources of Defects

What exactly causes chips to fail? The sources of yield loss are varied, but they all stem from the incredible precision required in semiconductor manufacturing. When you're working with features measured in nanometers (billionths of a meter), even the tiniest imperfection can cause catastrophic failure.

### 12.2.1 Particle Defects

The most common yield killers are **particles**, microscopic bits of dust, metal, plastic, or other materials that interfere with the manufacturing process. To understand their impact, consider this perspective.

A typical human hair is about 75–100 μm (75,000–100,000 nm) thick. Modern transistor features can be <10 nm. This means a particle the size of a human hair is approximately 7500 times larger than some of the features we're trying to create! It's like trying to paint a detailed miniature while someone drops bowling balls on your canvas (Fig. 12.3).

Particles cause different problems depending on where they land:

- A particle on a transistor gate might create a short circuit
- A particle in a metal interconnect layer could break a connection
- A particle during photolithography can block light, preventing proper pattern transfer

A famous analogy in the industry compares a particle landing on a chip to a boulder falling on a highway. Just as the boulder blocks traffic, the particle disrupts the flow of electricity through carefully designed circuits.

**Fig. 12.3** Particle size comparison

### 12.2.2 Process Variation Defects

Not all defects are caused by foreign particles. Some result from variations in the manufacturing process itself:

- **Film thickness variations**: If deposited layers are too thick or too thin in certain areas
- **Etch problems**: Under-etching leaves material that should be removed; over-etching removes material that should stay
- **Alignment errors**: When one layer isn't perfectly aligned with the layer below it
- **Chemical contamination**: Impurities in processing chemicals or gases

These **process variations** might be subtle, perhaps a film that should be precisely 10 nm thick is 11 nm in one area of the wafer, but even small deviations can cause chips to fail.

### 12.2.3 Equipment-Related Defects

The complex tools used in semiconductor manufacturing can themselves introduce defects:

- A malfunctioning robot arm might scratch a wafer
- A worn component might shed particles
- A misaligned chamber part might cause uneven processing

- A temperature controller that's slightly off could affect reaction rates

One infamous real-world example occurred at a major fab where mysterious defects appeared suddenly on wafers. After extensive investigation, engineers discovered that a tiny rubber O-ring in a wafer handling system had degraded and was shedding microscopic rubber particles onto wafers. Replacing this simple component, which cost less than a dollar, solved a problem that had been causing millions in yield loss!

### 12.2.4 Wafer Edge Defects

The edges of wafers are particularly prone to defects for several reasons:

- They're handled more during transportation
- Some processes apply chemicals that can pool at the edges
- Edge regions may experience different temperatures during heating steps

For this reason, the outermost chips on a wafer typically have lower yields than those in the center.

### 12.2.5 Defect Density and Chip Size

An important concept in yield management is "**defect density**"the number of defects per unit area. As chips get larger, they're more likely to encompass a defect, reducing yield. This creates a fundamental tension in chip design. Larger chips can include more functionality but are more vulnerable to defects.

This relationship is formalized in what is known as the **Poisson yield model**:

$$Y = e^{-DA}$$

Where:

- $Y$ is the yield
- $D$ is the defect density (defects per unit area)
- $A$ is the area of the chip
- $e$ is the mathematical constant (approximately 2.718)

This exponential relationship means that doubling chip size can dramatically decrease yield if defect density remains constant. This explains why many large, complex chips are now being designed as multiple smaller "chiplets" that are packaged together. Smaller dies have inherently better yield.

### 12.2.6 The Shrinking Defect Challenge

As semiconductor feature sizes continue to shrink, the size of a "**killer defect**" (one large enough to cause failure) also shrinks. This means that particles that would have been harmless in older technologies can now cause chips to fail.

This creates a seemingly impossible challenge: each new technology node requires even cleaner manufacturing environments and more precise processes than the previous generation. Meeting this challenge has driven the development of increasingly sophisticated cleanrooms, process controls, and inspection methods.

## 12.3 Process Control and SPC

Given the myriad ways defects can occur, how do semiconductor manufacturers ensure their processes consistently produce good chips? The answer lies in rigorous **process control** methods, particularly a technique called Statistical Process Control (SPC).

### 12.3.1 The Concept of Process Control

Process control is exactly what it sounds like, maintaining tight control over every aspect of the manufacturing process to ensure consistency and quality. In semiconductor manufacturing, this means monitoring and controlling hundreds of parameters, including:

- Gas flows and pressures
- Temperatures and heating rates
- Chemical concentrations
- Process timing
- Equipment settings

The goal is to keep all these parameters within their specified ranges or "specs" as these conditions are known to produce good results.

### 12.3.2 Statistical Process Control (SPC)

**Statistical Process Control (SPC)** is a method of monitoring a process using statistical techniques to identify when a process is drifting away from its target before it produces defective products. It's like having an early warning system that alerts you to potential problems before they cause yield loss.

The core tool of SPC is the **control chart**, a graph that plots measurements of a process parameter over time, typically with control limits that indicate acceptable variation (Fig. 12.4).

To understand how this works, consider a simple analogy of baking bread. If you know your bread turns out best when baked at 350 °F for 30 min, you could simply set your oven and timer accordingly. But what if your oven's temperature fluctuates? What if it's actually running at 375 °F when it says 350 °F?

An SPC approach would involve regularly measuring the oven's actual temperature and plotting it on a chart. If you notice the temperature is gradually rising over time, you can adjust it before your bread gets burned because it hasn't failed yet.

In semiconductor manufacturing, SPC is applied to hundreds of parameters. For example:

- The thickness of deposited films is measured after deposition steps
- Critical dimensions (feature sizes) are measured after lithography and etching
- Electrical parameters are tested at various stages

### 12.3.3 Control Limits and Rules

SPC charts typically include:

- A center line representing the target value
- Upper and lower **control limits** (usually set at ±3 standard deviations)
- The actual data points collected over time

Engineers use various rules to detect when a process might be drifting out of control:

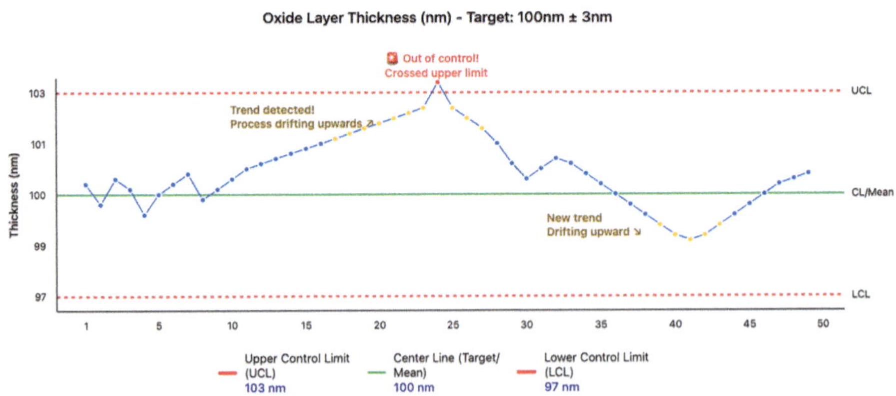


**Fig. 12.4** Statistical process control chart

- Points outside the control limits
- A series of consecutive points all above or below the center line
- A trend of steadily increasing or decreasing values
- A pattern of abnormal variation

When these signals appear, action is taken before the process produces defective chips.

### 12.3.4 Test Wafers and Monitoring

To implement SPC effectively, fabs regularly process special "**test wafers**" or "**monitor wafers**" that are measured but not intended to become products. These provide data on process performance without risking actual product wafers.

For example, before running a batch of product wafers through a critical deposition tool, an engineer might first run a dummy lot with single or multiple test wafers to verify that the tool is depositing films of the correct thickness. Only if the test wafer results look good will the product wafers be processed.

### 12.3.5 Design of Experiments (DOE)

While SPC helps maintain existing processes, improving those processes often involves another statistical technique: **Design of Experiments (DOE)**. DOE is a systematic method for determining the relationship between the factors affecting a process and the output of that process.

Rather than changing one variable at a time, DOE evaluates multiple variables simultaneously through carefully designed experimental runs. This allows engineers to efficiently identify which factors most strongly influence the results and find the optimal settings.

For example, a DOE might vary deposition temperature, gas flow rates, and pressure across a series of test wafers to find the combination that produces the most uniform film with the fewest defects.

### 12.3.6 The Role of Metrology

Critical to all process control is **metrology**, the science of measurement. Semiconductor manufacturing employs an incredible array of measurement technologies.

- Optical microscopes for visual inspection
- Scanning electron microscopes for nano-scale imaging
- Atomic force microscopes for 3D surface mapping
- X-ray techniques for analyzing materials

- Electrical testing for device performance

These tools provide the data that feeds into SPC systems, allowing engineers to detect and correct problems before they impact yield.

## 12.4 Industry Practices and Anecdotes

The principles and methods we've discussed so far aren't just theoretical; they're put into practice every day in semiconductor fabs worldwide. Let's look at some real-world examples of how the industry approaches yield management.

### 12.4.1 Intel's "Copy Exactly!" Methodology

One of the most famous yield management strategies in the industry is Intel's "**Copy Exactly!**" philosophy, developed in the 1980s. The premise is simple but powerful. Once a manufacturing process is working well in one fab, it should be copied exactly, down to the tiniest detail, when transferred to another fab.

This means identical equipment, identical settings, identical chemical suppliers, identical maintenance procedures, and even identical facility details like air handling systems. Nothing changes without careful evaluation.

The story goes that during one fab transfer, engineers were required to use the exact same length and type of vacuum hose between certain pieces of equipment, even though there seemed to be no technical reason why a slightly different hose would matter. Later analysis revealed that subtle vibrations transmitted through the hose actually did impact process results!

This extreme attention to detail might seem excessive, but it stems from a fundamental truth about semiconductor manufacturing: the processes are so complex that we don't always fully understand all the factors that influence results. By copying everything exactly, companies avoid introducing unknown variables that might reduce yield.

### 12.4.2 The Yield Detective Story

Yield improvement often resembles detective work, requiring persistence and attention to detail. One classic industry story involves a fab experiencing mysterious, intermittent defects that appeared as tiny holes in metal lines on certain wafers.

The yield team tried everything. They checked for particle sources, examined chemicals for contamination, verified equipment settings, and replaced components. All to no avail. The defects continued to appear randomly.

Finally, an engineer noticed a subtle pattern: the affected wafers tended to be processed on Mondays and after holidays. This led them to investigate the idle time of equipment, which revealed that when a particular chamber sat unused over weekends, a small amount of moisture accumulated. This moisture then caused microscopic defects when processing resumed.

The solution was simple: a modified "idle recipe" that periodically heated the chamber during downtime to prevent moisture buildup. Yield immediately improved, saving millions of dollars.

This story illustrates how yield improvement often requires cross-functional teams looking at problems from multiple angles, with process engineers, equipment engineers, quality specialists, and statisticians working together to solve complex puzzles.

### 12.4.3 Yield Learning Curve

When a new manufacturing process is introduced, it typically follows what's called a "**yield learning curve**" starting with relatively low yields that improve as engineers identify and fix issues. This learning process is institutionalized in the industry through methodical approaches.

1. **Data collection**: Gathering comprehensive information about defects and process parameters
2. **Pareto analysis**: Identifying the most common or impactful defect types to prioritize
3. **Root cause analysis**: Investigating the underlying causes of priority defects
4. **Corrective action**: Implementing solutions to address root causes
5. **Verification**: Confirming that the solutions actually improve yield
6. **Standardization**: Incorporating successful solutions into standard procedures

This cycle repeats continuously, with teams tackling the highest-impact issues first, then moving to less common problems as yields improve (Fig. 12.5).

### 12.4.4 Designed for Manufacturability (DFM)

Modern yield management begins before manufacturing starts, through an approach called **Design for Manufacturability (DFM)**. Chip designers work with manufacturing engineers to create designs that are inherently more tolerant of process variations and less likely to be affected by defects.

DFM techniques include:

- Avoiding minimum-sized features where possible
- Adding redundant vias between metal layers

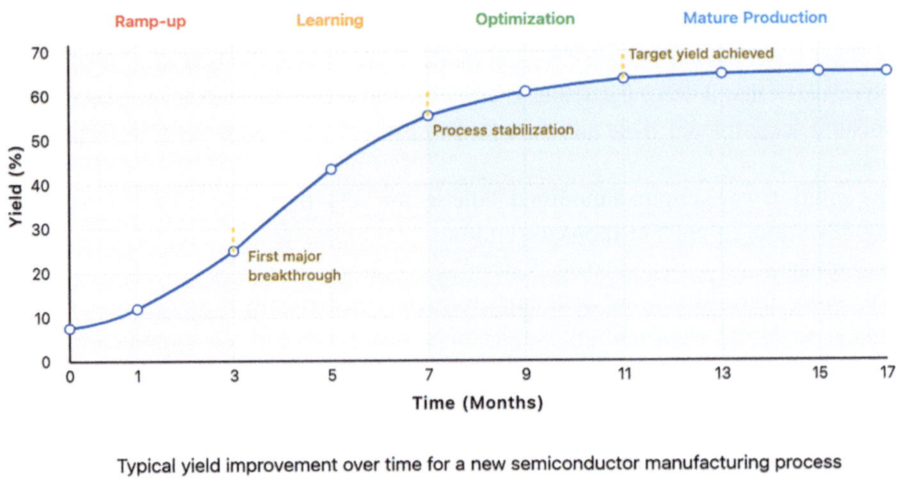


**Fig. 12.5** Yield learning curve. Note: This yield learning curve represents a typical Yield learning curve scenario for illustration purposes; actual yield rates, improvement trajectories, and timelines vary significantly based on technology node, product type, and manufacturing complexity

- Implementing uniform pattern densities to improve process consistency
- Including test structures that help diagnose problems

Some advanced chips even include redundant circuits that can be activated if the primary circuit has a defect. Memory chips, for example, often include spare rows and columns that can replace defective ones, significantly improving yield.

### 12.4.5 Inline Inspection vs. End-of-Line Testing

Yield management involves a balance between:

- **Inline inspection**: Checking wafers during processing to catch problems early
- **End-of-line testing**: Comprehensive testing of completed chips

Both are essential. Inline inspection helps prevent wasting further processing on wafers with known defects, while end-of-line testing ultimately determines which chips meet specifications.

Modern fabs employ automated optical inspection systems that can rapidly scan wafers between process steps, identifying defects as small as a few nanometers. These systems compare images to known good patterns or to adjacent identical chips, flagging any differences for review.

When defects are found, engineers must decide whether to continue processing the wafer (if the defects are minor or in non-critical areas) or to scrap it (if the defects are severe). This decision balances the cost of further processing against the likelihood of producing usable chips.

### 12.4.6 The Human Element

For all the advanced technology and statistical methods, yield management still has a critical human component. Experienced engineers develop an almost intuitive feel for their processes, noticing subtle changes that automated systems might miss.

Fab culture plays a role too. The best-performing fabs tend to have cultures that emphasize:

- Rigorous adherence to procedures
- Open communication about problems
- Collaborative problem-solving
- Recognition for identifying and solving issues
- A sense of ownership and pride in quality

As one industry veteran put it, "Robots don't improve yield, people do."

**Summary**

- Yield, the percentage of good chips produced, is critical to semiconductor manufacturing economics. Even small yield improvements can dramatically reduce the cost per chip and increase profitability.
- Defects arise from many sources, including particles, process variations, and equipment issues. As feature sizes shrink, controlling these defects becomes increasingly challenging, requiring exceptionally clean environments and precise processes.
- Statistical Process Control (SPC) enables manufacturers to detect when processes drift away from optimal settings before they produce defective chips. By continuously monitoring key parameters and using control charts, engineers can maintain consistent production.
- Successful yield management combines rigorous methodologies like "Copy Exactly!" with systematic problem-solving approaches. Yield improvement is a continuous process of data collection, analysis, experimentation, and implementation of solutions.
- The evolution of semiconductor manufacturing has been as much about mastering process control as it has been about technological innovation. Today's high yields result from decades of accumulated knowledge and increasingly sophisticated control techniques.

## Review Questions

1. If a wafer contains 600 dies and 480 pass final testing, what is the die yield? If each wafer costs $6000 to process, what is the cost per good die? How would the cost per die change if yield improved to 90%?
2. Explain why defect density becomes more critical as chip feature sizes shrink. Use an analogy to illustrate your explanation.
3. What is Statistical Process Control (SPC) and why is it important in semiconductor manufacturing? Describe a scenario where SPC might detect a problem before it affects yield.
4. Imagine you're a yield engineer and you notice that wafers processed on a particular tool have lower yield than identical wafers processed on other tools. Outline the steps you would take to investigate and address this issue.
5. Research a real-world example of a semiconductor yield or quality problem, such as the Intel Pentium FDIV bug or the AMD TLB bug. Summarize what happened and what lessons the industry learned from the incident.
6. How might the principles of yield management and process control that you've learned in this chapter apply to other industries or even everyday life situations?

# 13 Advanced Node Challenges and Future Trends: Pushing the Limits

Throughout this book, we've explored the remarkable journey of how a humble grain of sand is transformed into one of humanity's most sophisticated creations, the modern semiconductor chip. We've seen how silicon is purified, sliced into wafers, patterned with intricate designs, etched with precision, and augmented with carefully placed impurities to create billions of transistors and connections.

But where does the semiconductor industry go from here? After decades of relentless miniaturization and performance improvements, chipmakers are now pushing against the fundamental limits of physics. In this final chapter, we'll explore the cutting-edge challenges confronting the industry and the innovative solutions being developed to overcome them.

We'll examine why traditional scaling is becoming increasingly difficult, the revolutionary new lithography and transistor designs enabling continued progress, and the alternative approaches that may define the future of computing. We'll also look beyond pure technology to consider the global ecosystem that makes this industry possible.

## 13.1 The End of Easy Scaling (Moore's Law)

For over half a century, the semiconductor industry has lived by a remarkable prediction made by Intel co-founder Gordon Moore in 1965. Moore observed that the number of transistors on an integrated circuit doubled approximately every 2 years, enabling dramatic improvements in computing power at ever-lower costs.

This prediction, known as "**Moore's Law**," wasn't a physical law but rather an observation that became a self-fulfilling prophecy. The entire industry aligned its research, development, and manufacturing around this cadence of improvement, driving an unprecedented period of technological advancement.



A. D. Watpade, *From Sand to Chip*, Synthesis Lectures on Engineering, Science, and Technology, https://doi.org/10.1007/978-3-032-18439-9_13

### 13.1.1 Hitting Physical Barriers

Imagine you're trying to draw increasingly detailed pictures in a small box. At first, using a finer-tipped pen lets you add more details. But eventually, you reach a point where the pen tip is approaching the size of individual ink molecules. No matter how steady your hand or how precise your tools, you simply can't draw lines smaller than the molecules that make them up.

This analogy captures the fundamental challenge facing semiconductor manufacturers today. When transistors were measured in micrometers (millionths of a meter), shrinking them was primarily an engineering challenge, difficult but achievable with better tools and techniques. Now that we're approaching features measured in single-digit nanometers (billionths of a meter), we're confronting the atomic nature of matter itself.

At these scales, several fundamental physical issues emerge:

- **Quantum Tunneling**: Electrons, being quantum particles, don't always behave like tiny balls. They can sometimes "tunnel" through barriers that classical physics says they shouldn't be able to penetrate. As transistor gates get thinner, electrons can tunnel through them, causing current leakage and preventing the transistor from fully turning off.
- **Heat Dissipation**: Smaller transistors packed more densely generate tremendous heat in a tiny area. Modern high-performance chips can easily reach temperatures that would damage them if not actively cooled.
- **Variability**: When features are just a few atoms wide, even minute variations in manufacturing can lead to significant differences in transistor behavior.
- **Interconnect Delays**: As transistors get smaller, the metal wires connecting them don't necessarily shrink proportionally. The delay caused by signals traveling through these interconnects becomes a limiting factor in performance.

### 13.1.2 Economic Challenges

Beyond these physical barriers, economic factors are also slowing Moore's Law:

- **Skyrocketing Fab Costs**: Each new technology node requires more sophisticated equipment and facilities. A leading-edge fab now costs upwards of $20 billion to build, a staggering investment even for the largest semiconductor companies.
- **Diminishing Returns**: The performance improvements gained from each new node are becoming smaller while costing more to achieve.
- **Design Complexity**: Designing for advanced nodes requires increasingly sophisticated tools and larger engineering teams, raising the cost of chip development.

### 13.1.3 Beyond Simple Scaling

Despite these challenges, the semiconductor industry continues to advance, not just by shrinking transistors, but by finding new ways to improve performance and functionality. As the former CEO of Intel, Brian Krzanich, once noted, "Moore's Law is fundamentally a law of economics, and when we say Moore's Law is alive and well, what we're really saying is that the fundamental economics of the industry are alive and well."

The future of chip advancement now relies on multiple strategies working in concert:

- New transistor architectures that improve performance without necessitating proportional size reductions
- Advanced packaging techniques that combine multiple chips into systems
- Specialized designs optimized for specific applications rather than general-purpose computing
- Novel materials that enable better performance at the same dimensions

These approaches reflect a shift from "scaling down" to "**scaling out and up**" finding new ways to arrange and connect transistors rather than just making them smaller.

## 13.2 Extreme Lithography (EUV and Beyond)

Even as traditional scaling becomes more challenging, the industry continues to push the boundaries of what's possible in lithography, the critical process that defines the smallest features on a chip.

### 13.2.1 The EUV Revolution

For decades, chip manufacturers relied on argon fluoride (ArF) immersion lithography, using light with a wavelength of 193 nm to print features much smaller than the wavelength itself through various optical tricks. But this approach eventually reached its practical limits.

The solution emerged in the form of **Extreme Ultraviolet (EUV) lithography**, which uses light with a wavelength of just 13.5 nm, about 14 times shorter than ArF light. This shorter wavelength allows for much finer patterns to be printed directly, enabling the production of chips at the 7 nm node and beyond (Fig. 13.1).

EUV represents one of the most complex technologies ever developed. To understand just how remarkable it is, consider some of its key elements:

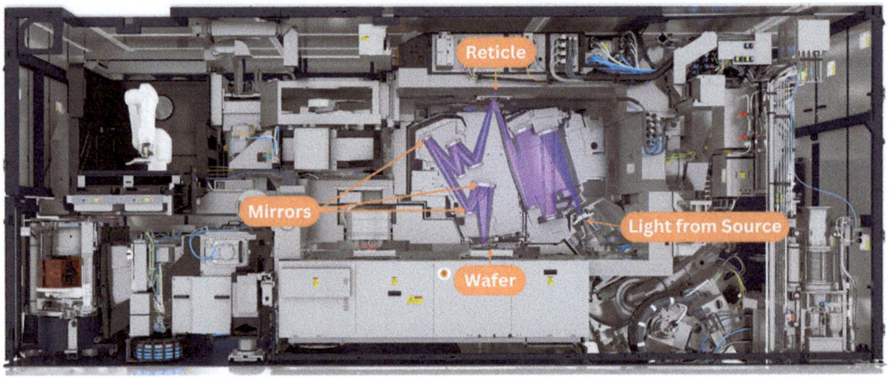


**Fig. 13.1** EUV lithography system. Credit: ASML

- **Generating EUV Light**: EUV light is created by firing high-powered lasers at tiny droplets of molten tin. When the laser hits a tin droplet, it turns the tin into plasma that emits EUV light. This happens 50,000 times per second with micron-precision aiming.
- **All-Mirror Optics**: EUV light is absorbed by virtually all materials, including the lenses used in traditional lithography. Instead, EUV systems use complex mirror systems with multiple layers of alternating materials that can reflect about 70 percent of the EUV light, but this means that after bouncing off several mirrors only a small fraction of the original light reaches the wafer.
- **Perfect Vacuum**: EUV light is absorbed by air, so the entire system must operate in an ultra-high vacuum environment.
- **Mask Technology**: The photomasks (templates) for EUV are not transparent but reflective, requiring entirely new fabrication and inspection techniques.

These technical challenges made EUV development a decades-long journey. The first commercial EUV systems, produced by ASML in the Netherlands, cost roughly $150 million each and are among the most complex machines ever built. Despite their cost, leading-edge manufacturers including TSMC, Samsung, and Intel have invested heavily in EUV technology to enable their most advanced processes.

### 13.2.2 Beyond EUV

Even as EUV becomes established for leading-edge manufacturing, researchers are already exploring what comes next:

- **High-NA EUV**: Increasing the numerical aperture (NA) of EUV systems from 0.33 to 0.55 to enable even finer resolution.

- **Multiple Patterning**: Using multiple exposures to create patterns more detailed than what a single exposure can achieve.
- **Direct-Write Technologies**: Electron-beam systems that "write" patterns directly onto wafers without using a mask, potentially useful for low-volume production of highly specialized chips.
- **Nanoimprint Lithography**: A technique that physically imprints patterns onto a substrate, similar to how a printing press works, but at the nanoscale.

While each of these approaches has advantages for specific applications, they also face significant technical hurdles before becoming viable for high-volume manufacturing. The quest for better lithography continues to be one of the semiconductor industry's greatest challenges and triumphs.

## 13.3 Next-Gen Transistors (FinFETs and Gate-All-Around)

Just as lithography has evolved to create smaller features, the fundamental design of the transistor has also undergone revolutionary changes to maintain performance at increasingly tiny dimensions.

### 13.3.1 The Evolution from Planar to 3D

For decades, transistors used a **planar** (flat) design. The gate electrode sat atop a flat channel, controlling the flow of current from source to drain. This design worked well until transistors shrank to around 28–22 nm. At that point, the control exerted by the gate became insufficient, leading to leakage currents and higher power consumption.

The industry's answer was the **FinFET**, a three-dimensional transistor architecture first introduced to volume manufacturing by Intel in 2011. In a FinFET, the channel is formed into a thin vertical fin that rises above the substrate. The gate wraps around three sides of this fin, providing much better control over the current flow (Fig. 13.2).

To understand the advantage of this design, imagine trying to control the flow of water through a garden hose. With a planar transistor, you're pressing down on the hose from just one side. With a FinFET, you're gripping the hose on three sides, giving you much better control over the water flow.

FinFETs enabled the continuation of Moore's Law scaling for several more generations, from 22 nm down to 3 nm. However, as dimensions continued to shrink, even the three-sided control of FinFETs began to reach its limits.

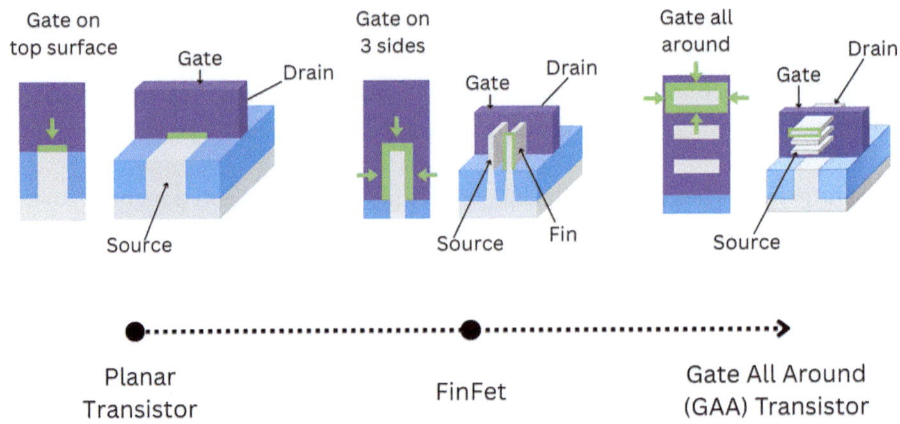


**Fig. 13.2** Transistor architecture evolution: Planar vs. FinFET vs. Gate-All-Around structures

### 13.3.2 Gate-All-Around: The Next Evolution

The latest evolution in transistor design is the **gate-all-around (GAA)** transistor, sometimes called **nanosheet** or **nanowire** transistors depending on their specific implementation. In this design, the channel is completely surrounded by the gate on all sides, providing optimal control over current flow.

Continuing our water hose analogy, a gate-all-around transistor is like completely encircling the hose, giving maximum control over the water flowing through it.

Samsung introduced the first commercial gate-all-around transistors in 2022 with their 3 nm process, with other manufacturers following suit. These new transistors offer better performance, lower power consumption, and continued scaling benefits compared to FinFETs.

### 13.3.3 The Impact on Manufacturing

These architectural changes have profoundly affected manufacturing processes. Each new transistor design requires not just new lithography techniques, but also new deposition, etching, and doping approaches. The complexity of manufacturing has increased dramatically, as has the precision required at each step.

For example, creating the suspended nanosheets in a GAA transistor requires carefully etching away sacrificial material between the sheets while preserving the delicate structure. This level of manufacturing precision was unimaginable just a few decades ago.

Despite these challenges, the evolution of transistor design has allowed the industry to continue improving performance even as traditional dimensional scaling slows down. By moving into the third dimension and finding clever new architectures, engineers have found ways to squeeze more performance out of every square nanometer of silicon.

## 13.4 Power and Thermal Limits

While much of the focus in semiconductor advancement has been on performance and density, another challenge has emerged as equally critical: power consumption and the resulting heat generation.

### 13.4.1 The Power Wall

Imagine a string of light bulbs. If you keep adding more and more bulbs to the string while keeping the same wire, eventually you'll either blow a fuse or the wire will overheat. Similarly, as we pack more transistors into the same area, the chip's power density increases dramatically.

This creates what engineers call the "**power wall**" which is a fundamental limit to how much power can be consumed in a given area before the chip becomes too hot to function reliably. Modern high-performance processors can easily generate more than 100 W of heat in an area smaller than a postage stamp. Without adequate cooling, such a chip would quickly reach temperatures that damage the silicon (Fig. 13.3).

The power problem manifests differently depending on the application:

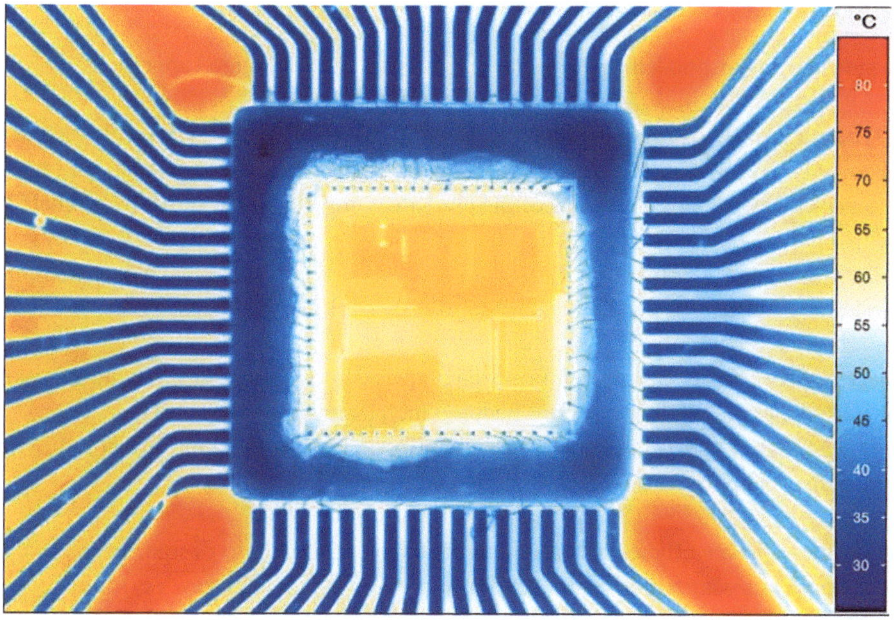


**Fig. 13.3** Thermal map of a modern processor. Credit: InfraTec GmbH

- **Mobile Devices**: Battery life is the primary constraint. A smartphone processor that's too power-hungry will drain the battery quickly, regardless of its performance.
- **Data Centers**: Power consumption translates directly to electricity costs and cooling requirements. A more efficient chip saves money in ongoing operational expenses.
- **High-Performance Computing**: Cooling becomes the limiting factor. Even with advanced cooling systems, there's a practical limit to how much heat can be removed from a chip.

### 13.4.2 The End of Dennard Scaling

The power challenge was exacerbated by the breakdown of **Dennard scaling**, a principle that previously helped manage power consumption. Dennard scaling held that as transistors got smaller, their power density remained constant because voltage and current scaled down with size.

Around 2005, this principle began to fail as transistors became so small that their supply voltages couldn't be reduced further without affecting reliability. As a result, power density began to increase with each generation, leading to what is known as "**dark silicon**" portions of a chip that must remain powered down most of the time to avoid overheating.

### 13.4.3 Managing Power and Heat

The industry has responded to these challenges with a multi-faceted approach:

- **Dynamic Power Management**: Modern chips constantly adjust their voltage and frequency based on workload, reducing power when maximum performance isn't needed.
- **Power Gating**: Unused sections of the chip are completely turned off to eliminate both active power and leakage.
- **Heterogeneous Computing**: Combining different types of cores (high-performance and energy-efficient) on the same chip allows workloads to be assigned to the most appropriate processor.
- **Advanced Cooling Solutions**: From heat spreaders and heat pipes to liquid cooling systems, thermal management has become increasingly sophisticated.
- **3D Packaging with Thermal Considerations**: New packaging technologies must carefully manage heat flow between stacked dies.

The focus on power efficiency has fundamentally changed chip design. In many cases, performance per watt has become a more important metric than raw performance. This shift has driven interest in specialized processors that deliver high efficiency for specific workloads instead of general-purpose computing, a trend we'll explore further in later sections.

## 13.5 Heterogeneous Integration (Chiplets and 3D Stacking)

As traditional scaling faces increasing challenges, the industry has found a new direction: instead of making individual chips larger and more complex, why not combine multiple smaller chips into more powerful systems?

This approach, known as **heterogeneous integration**, has rapidly emerged as one of the most promising paths forward for semiconductor advancement.

### 13.5.1 The Chiplet Revolution

A **chiplet** is essentially a small, specialized chip designed to be combined with other chiplets in a single package. Instead of creating one massive, complex die that does everything, designers can mix and match chiplets with different functions, connecting them with high-speed links (Fig. 13.4).

This modular approach offers several key advantages:

- **Improved Yield**: Smaller dies have higher manufacturing yields than larger ones. By breaking a large design into multiple smaller chiplets, manufacturers can significantly reduce costs.

**Fig. 13.4** Chiplet-based processor architecture. Example: Intel Xeon 6 processor showing external package (left) and internal chiplet layout (right). One package containing multiple chiplets working together. Credit: Intel Corporation

- **Process Optimization**: Different components can be manufactured using the process best suited for each function. Memory, analog functions, and digital logic all have different ideal manufacturing processes.
- **Mix-and-Match Flexibility**: Chiplets enable companies to reuse components across multiple products, reducing design costs and time-to-market.

Several major companies have embraced the chiplet approach. AMD's Ryzen processors use chiplets to combine multiple CPU cores manufactured on advanced processes with I/O components made on more mature processes. Intel uses similar strategies with their Foveros technology.

### 13.5.2 2.5D and 3D Integration

Beyond simply placing chiplets side-by-side, the industry is exploring ways to stack and integrate chips in three dimensions:

**2.5D Integration**: In this approach, multiple chips are mounted on a **silicon interposer**, essentially a passive piece of silicon with microscopic wiring that connects the chips. The interposer allows for thousands of connections between chips, much more than traditional packaging would permit. High-bandwidth memory (HBM) used with graphics processors is a prominent example of 2.5D integration.

**3D Stacking**: Even more advanced is true **3D integration**, where chips are literally stacked on top of each other and connected with **through-silicon vias (TSVs)** that are vertical connections that pass completely through the silicon. This approach offers the highest interconnect density and shortest connection lengths but also presents challenges for power delivery and heat dissipation. (Refer to Chap. 11 for 2.5D and 3D integration and stacking).

To visualize 3D stacking, imagine building a multi-story building instead of spreading out over more land. Just as cities build upward when land is scarce, chip designers are building vertically when planar space is at a premium.

### 13.5.3 Advanced Packaging Technologies

Making heterogeneous integration possible requires sophisticated packaging technologies:

- **Advanced Substrates**: Special substrates with fine-pitch wiring to connect chiplets.
- **Micro-Bumps**: Tiny solder connections between stacked dies or between dies and interposers.
- **Through-Silicon Vias (TSVs)**: Vertical connections passing through silicon dies to connect stacked chips.
- **Thermal Management**: Special materials and designs to manage heat in stacked structures.

As former AMD CEO Lisa Su noted, "Packaging is the new Moore's Law." While transistors may not be shrinking as quickly as before, the way we connect them continues to evolve rapidly, enabling continued performance improvements through system-level innovation rather than just device scaling.

## 13.6 Specialized and AI Chips

Another major trend in semiconductor evolution is the shift from general-purpose processors to specialized chips designed for specific workloads. This specialization allows for dramatically improved performance and efficiency by optimizing the architecture for particular tasks.

### 13.6.1 The Rise of Domain-Specific Architectures

For decades, general-purpose CPUs dominated computing, with their architecture optimized for sequential processing of diverse workloads. While versatile, this approach is inherently inefficient for many specific tasks.

Imagine a multi-purpose tool versus specialized tools. A Swiss Army knife can handle many tasks adequately, but a dedicated screwdriver will always be better at turning screws. Similarly, specialized processors can dramatically outperform general-purpose CPUs for specific workloads.

This specialization trend began with **Graphics Processing Units (GPUs)**, which were optimized for the parallel calculations needed in 3D rendering. When researchers realized these same parallel architectures were ideal for certain types of machine learning calculations, GPUs found a new role in accelerating AI workloads.

### 13.6.2 AI Accelerators

The explosive growth of artificial intelligence, particularly deep learning, has driven demand for chips specifically designed to handle neural network computations (Fig. 13.5). These **AI accelerators** feature:

- **Massive Parallelism**: Thousands of simple processing elements working simultaneously.
- **Optimized Data Flow**: Architectures designed to move data efficiently for matrix operations.

**Fig. 13.5** AI accelerator architecture. Example: Intel's Gaudi chip, a specialized AI processor designed with dedicated circuits to handle neural network calculations. Its central processor and high-bandwidth memory chips work together to speed up deep learning tasks more effectively than general-purpose processors. Credit: Intel Corporation

- **Reduced Precision**: Many AI workloads can use lower precision (8-bit or even 4-bit) calculations instead of standard 32-bit or 64-bit math, allowing for more compute units in the same area.
- **Specialized Memory Hierarchies**: Custom memory systems designed for the access patterns of neural networks.

Examples of specialized AI chips include:

- **Google's Tensor Processing Units (TPUs)**: Custom-designed for Google's TensorFlow machine learning framework.
- **NVIDIA's Tensor Cores**: Specialized units within their GPUs specifically for AI workloads.
- **Apple's Neural Engine**: An AI accelerator in iPhone and iPad processors for tasks like face recognition and natural language processing.
- **Cerebras Wafer-Scale Engine**: An extreme example that uses almost an entire wafer as a single AI chip.

These specialized chips can achieve orders of magnitude better performance per watt compared to general-purpose processors for their target workloads.

### 13.6.3 Beyond AI: Domain-Specific Acceleration

The specialization trend extends beyond just AI:

- **Network Processors**: Optimized for packet processing in network equipment.
- **Storage Controllers**: Specialized for managing flash memory in SSDs.
- **Security Accelerators**: Dedicated to encryption and cryptographic operations.
- **Signal Processing Chips**: Optimized for audio, video, and sensor data processing.

This diversification represents a fundamental shift in computing architecture. Rather than relying on one processor type to handle all tasks, systems increasingly use a heterogeneous mix of specialized processors working together, each handling the workloads it's best suited for.

## 13.7 Emerging Materials and Devices

While silicon CMOS technology continues to advance, researchers are exploring entirely new materials and device concepts that could eventually supplement or even replace traditional semiconductors in certain applications.

### 13.7.1 2D Materials

One of the most exciting areas of research involves **two-dimensional (2D) materials**, crystalline materials just one or a few atoms thick. The most famous is **graphene**, a single-atom-thick layer of carbon arranged in a honeycomb pattern.

Graphene has remarkable properties, such as being incredibly strong, highly conductive, and transparent. However, it lacks a natural bandgap (the energy difference that allows semiconductors to switch on and off), limiting its use in transistors without modification.

Other 2D materials have emerged to address this limitation:

- **Transition Metal Dichalcogenides (TMDs)**: Materials like molybdenum disulfide ($MoS_2$) that have natural bandgaps suitable for transistors.
- **Hexagonal Boron Nitride (hBN)**: An excellent insulator that can be used as a substrate or gate dielectric for other 2D materials.
- **Black Phosphorus**: A semiconductor with tunable properties depending on its thickness.

The appeal of these materials lies in their ultimate thinness. A transistor made from a truly 2D material could be just a few atoms thick, potentially allowing for even higher density than silicon devices. Additionally, many 2D materials offer unique properties like flexibility, transparency, or special optical characteristics that could enable entirely new applications.

### 13.7.2 Photonic Computing

Another promising frontier is **photonic computing**, using light instead of electricity to process information. Photonic components can transmit data with less energy and higher bandwidth than electrical connections.

Researchers are developing:

- **Silicon Photonics**: Integrating optical components with traditional silicon electronics.
- **Photonic Integrated Circuits**: Chips that process light signals directly.
- **Optical Interconnects**: Using light to transmit data between chips or within a chip.

While fully photonic computers remain distant, hybrid electronic-photonic systems are already emerging, particularly for data-intensive applications like AI and high-performance computing, where moving data consumes more energy than processing it.

### 13.7.3 Quantum Computing

Perhaps the most radical departure from conventional computing is **quantum computing**. Instead of using bits that are either 0 or 1, quantum computers use **quantum bits** or "**qubits**" that can exist in multiple states simultaneously due to quantum superposition.

Quantum computers aren't meant to replace classical computers for everyday tasks, but they could potentially solve certain problems that are practically impossible for conventional computers, such as simulating quantum systems or factoring very large numbers.

Several quantum technologies are being explored:

- **Superconducting Qubits**: Circuits cooled to near absolute zero, where quantum effects dominate.
- **Trapped Ions**: Individual atoms held in electromagnetic fields.
- **Topological Qubits**: Theoretical qubit types that would be inherently more stable.

While practical quantum computers with thousands of reliable qubits remain a future goal, the field is advancing rapidly, with potential to revolutionize areas like materials science, cryptography, and complex system optimization.

### 13.7.4 Neuromorphic Computing

Inspired by the brain's architecture, **neuromorphic computing** aims to create chips that more closely mimic biological neural systems. Unlike conventional digital systems that separate memory and processing, neuromorphic chips integrate the two, potentially offering massive efficiency improvements for certain tasks.

Technologies enabling neuromorphic computing include:

- **Memristors**: Devices whose resistance changes based on the history of current that has flowed through them, similar to how neural synapses strengthen with use.
- **Spintronics**: Using electron spin rather than charge to store and process information.
- **Analog Computing Elements**: Components that perform computations in the analog domain rather than digitally.

While these emerging technologies may not replace mainstream CMOS in the near term, they represent exciting new directions that could lead to dramatic advances in specific application areas, complementing traditional semiconductor technology.

## 13.8 The Global Semiconductor Ecosystem

The semiconductor industry is remarkable not just for its technological achievements but also for its global, interconnected nature. Understanding this ecosystem is crucial for appreciating the future of semiconductor manufacturing.

### 13.8.1 A Worldwide Supply Chain

Modern semiconductor production involves one of the most complex supply chains of any industry, spanning numerous countries and thousands of companies:

- **Design**: Often happens in the US (Silicon Valley, Austin, Boston), Europe, Israel, and increasingly in China and India.
- **Manufacturing Equipment**: Dominated by companies from the US (Applied Materials, Lam Research), Japan (Tokyo Electron), and the Netherlands (ASML).
- **Materials**: Specialty chemicals, gases, and silicon wafers come from suppliers in Japan, Germany, the US, Korea, and Taiwan.
- **Fabrication**: Concentrated in East Asia (Taiwan, South Korea, Japan), with significant capacity also in the US, Europe, and China.
- **Assembly and Testing**: Often performed in Southeast Asia (Malaysia, Philippines, Vietnam) and China.

This distributed supply chain has enabled incredible innovation and cost efficiency but also creates vulnerabilities when disrupted by geopolitical tensions, natural disasters, or pandemics.

### 13.8.2 The Rise of Foundries and the Fabless Model

One of the most significant industry developments has been the separation of design and manufacturing. Under this model:

- **Fabless Companies**: Focus solely on designing chips without owning manufacturing facilities (e.g., Qualcomm, NVIDIA, AMD, Apple).
- **Foundries**: Specialize in manufacturing chips designed by others (e.g., TSMC, Samsung, GlobalFoundries).

This division of labor has allowed for greater specialization and efficiency, but has also concentrated advanced manufacturing in a few companies and regions. Taiwan Semiconductor Manufacturing Company (TSMC) now produces over 90% of the world's most advanced chips, creating both a marvel of industrial efficiency and a potential strategic vulnerability.

### 13.8.3 National Initiatives and Reshoring

Recent years have seen growing recognition of semiconductors as strategically critical technology, leading to major government initiatives worldwide:

- **CHIPS Act (US)**: A $52 billion program to rebuild semiconductor manufacturing capacity in the United States.
- **European Chips Act**: The EU's €43 billion effort to double its share of global semiconductor production.
- **China's Semiconductor Initiatives**: Massive investments aimed at developing self-sufficiency in chip production.
- **South Korea's K-Semiconductor Strategy**: A public-private partnership to strengthen the country's leadership in memory and advanced logic.

These initiatives reflect a shift from treating semiconductors as merely commercial products to viewing them as strategic assets with national security implications.

### 13.8.4 Sustainability Challenges

As the industry continues to grow, sustainability concerns are becoming increasingly important:

- **Energy Usage**: Semiconductor fabs are among the most energy-intensive manufacturing facilities, with a single advanced fab consuming as much electricity as a small city.
- **Water Consumption**: Chip production requires enormous quantities of ultra-pure water, creating challenges in water-stressed regions.
- **Chemical Usage**: The industry uses numerous specialty chemicals, some with significant environmental footprints.
- **Electronic Waste**: The growing volume of electronic devices creates end-of-life disposal challenges.

Addressing these sustainability issues while continuing to advance technology will be a critical challenge for the industry's future.

### 13.8.5 Collaboration and Competition

Perhaps the most remarkable aspect of the semiconductor industry is how it balances intense competition with necessary collaboration. Companies fiercely compete in product markets while simultaneously cooperating on standards, research consortia, and industry initiatives.

This balance of competition driving innovation while collaboration solves common challenges has been key to the industry's success and will likely remain essential as it tackles the increasingly difficult technical and economic hurdles ahead.

### 13.8.6 Looking Forward: Your Role in the Semiconductor Future

As we conclude our journey from sand to chips, it's worth reflecting on the incredible human achievement that semiconductor manufacturing represents. In just a few decades, we've learned to manipulate matter at the atomic scale to create devices of staggering complexity that have transformed nearly every aspect of modern life.

The challenges ahead are formidable, but so is human ingenuity. The semiconductor industry has consistently overcome seemingly insurmountable barriers through creativity, collaboration, and persistence. There's every reason to believe this will continue.

For you, the reader, this knowledge opens doors to understanding and participating in this exciting field. Whether you're a student considering a career in semiconductors, a professional in an adjacent industry, or simply a curious mind wanting to understand how our digital world works, you now have insights into one of humanity's most remarkable technological achievements.

The future of semiconductor technology will be shaped by people who understand both its possibilities and its challenges. Perhaps you will be one of those who contributes to the next chapter in this extraordinary story.

## Summary

- Moore's Law is encountering fundamental physical and economic limits as transistors approach atomic scales, driving the industry to adopt new strategies beyond simple dimensional scaling.
- Advanced lithography using extreme ultraviolet (EUV) light and novel transistor architectures like FinFETs and gate-all-around designs have enabled continued progress at the smallest nodes, though at increasing cost and complexity.
- Power consumption and heat dissipation have become critical constraints, shifting focus from raw performance to energy efficiency and specialized designs.
- Heterogeneous integration through chiplets and 3D stacking provides new ways to improve system performance by combining specialized chips, effectively "scaling out" rather than just "scaling down."
- The future of semiconductors will likely involve a diverse ecosystem of technologies, including specialized AI accelerators, new materials beyond silicon, and potentially revolutionary approaches like quantum and neuromorphic computing, all within a complex global supply chain increasingly shaped by strategic national interests.

## Review Questions

1. Explain two fundamental physical or economic reasons why Moore's Law is becoming more difficult to sustain at advanced nodes. How are manufacturers adapting to this challenge?
2. What is EUV lithography, and why is it critical for manufacturing chips at the most advanced nodes? What are some of the technical challenges that had to be overcome to make EUV practical?
3. Why have specialized AI accelerator chips become so important in recent years? How do their architectures differ from general-purpose CPUs?
4. Describe one emerging technology or material that might play a role in future semiconductor devices. What advantages might it offer over traditional silicon CMOS technology?
5. (Research question) Choose one emerging computing approach (quantum, neuromorphic, photonic, etc.) and research its current state of development. What progress has been made recently, and what significant challenges remain before it could become commercially viable?

# Appendix: Resources and Further Reading

## Computing Milestones: Notable Chips and Developments

A brief selection of landmark chips in computing history, illustrating the progress in semiconductor technology.

| Year | Chip/milestone | Significance |
|---|---|---|
| 1947 | First Transistor | Invention of the point-contact transistor at Bell Labs. Ushered in the semiconductor era by replacing bulky vacuum tubes with a tiny solid-state device (though not a "chip," it's the essential beginning). |
| 1958 | First Integrated Circuit | Jack Kilby of Texas Instruments demonstrates the first ever integrated circuit (a simple oscillator) on a single piece of semiconductor. Proved that multiple components could be built in one device, laying the groundwork for microchips. |
| 1971 | Intel 4004 Microprocessor | The world's first commercial microprocessor (4-bit). Contained 2300 transistors and ran at ~740 kHz. Used in calculators, it demonstrated that an entire CPU could fit on one chip, kicking off the microprocessor revolution. |
| 1978 | Intel 8086 CPU | A 16-bit microprocessor that became the basis of the x86 architecture (the foundation of most PC processors). The 8086 (and its variant, the 8088 used in the IBM PC) helped standardize computer architecture. |
| 1989 | Intel 80486 CPU | A 32-bit microprocessor with an on-chip floating-point unit and over one million transistors. Showcased increasing integration (cache memory on-chip) and paved the way for the "Pentium" brand of CPUs. |
| 1993 | Intel Pentium | A 5th-generation x86 CPU (~3.1 million transistors). Introduced superscalar architecture (executing multiple instructions per clock) and became a household name via mass marketing, marking the era of rapidly accelerating PC performance. |
| 1999 | NVIDIA GeForce 256 GPU | Often dubbed the first "GPU" (Graphics Processing Unit). Brought parallel processing for graphics into the mainstream. Signaled the rising importance of specialized chips for specific tasks (in this case, 3D rendering). |



A. D. Watpade, *From Sand to Chip*, Synthesis Lectures on Engineering, Science, and Technology, https://doi.org/10.1007/978-3-032-18439-9

| Year | Chip/milestone | Significance |
|---|---|---|
| 2003 | AMD Opteron (K8) | A 64-bit x86 server/workstation processor. Introduced x86-64 (in tandem with AMD Athlon64 for desktops), extending the dominant x86 architecture to 64-bit computing, a crucial evolution for memory-intensive applications. |
| 2006 | Intel Core 2 Duo | A dual-core processor for consumers (two CPU cores on one chip). Marked the industry shift to multi-core designs as a way to improve performance within power limits, rather than solely increasing clock speed. |
| 2010 | Apple A4 SoC | Apple's first in-house system-on-chip for the iPhone/iPad. Integrated CPU, GPU, and other components on a single chip using ARM architecture. Demonstrated the growing trend of custom mobile SoCs optimized for power efficiency and multimedia. |
| 2017 | AMD Ryzen (Zen) CPU | A high-performance CPU that revived AMD's competitiveness. Notable for implementing a chiplet-based modular design in later iterations (Ryzen 3000 series in 2019) and for bringing high core counts to mainstream PCs, all on a power-efficient 7 nm process. |
| 2020 | Apple M1 SoC | A breakthrough ARM-based system-on-chip for by Apple for Apple laptops/desktops. Built on 5 nm technology with ~16 billion transistors, it combines an 8-core CPU, 8-core GPU, and neural engine. Achieved high performance with low power, sparking industry movement toward ARM in PCs and emphasizing tight hardware-software integration. |

This table highlights important chips that marked leaps in integration, architecture, or industry direction. They illustrate the trajectory from a single transistor to billions on a chip

## Recommended Books and Publications

- **Crystal Fire: The Birth of the Information Age** by Michael Riordan & Lillian Hoddeson. (W.W. Norton & Company, 1997.) A readable history of the transistor's invention and the dawn of electronics. Provides rich context about the people and science behind Bell Labs' breakthrough.
- **The Chip: How Two Americans Invented the Microchip and Launched a Revolution** by T.R. Reid. (Random House Trade Paperbacks, 2001.) An engaging account of Jack Kilby and Robert Noyce's independent inventions of the integrated circuit. Captures the early days of Silicon Valley and Texas Instruments.
- **Fabless: The Transformation of the Semiconductor Industry** by Daniel Nenni & Paul McLellan. (2014.) Explains the rise of the fabless business model and foundries. Good for understanding how companies like Qualcomm and Broadcom operate without their own fabs, and how TSMC changed the industry.
- **Semiconductor Manufacturing Handbook,** edited by Hwaiyu Geng. (2nd Ed., McGraw-Hill, 2017.) A comprehensive reference that covers modern semiconductor processes and equipment. It's more advanced, but useful for deeper dives into fabrication technology and industry best practices.
- **Introduction to Microelectronic Fabrication**, by Richard C. Jaeger. (2nd Ed., Prentice Hall, 2001.) A textbook that provides a step-by-step introduction to chip fabrication

processes. Though a bit dated regarding the latest nodes, it's excellent for fundamentals and is written in an accessible style for newcomers.

All the above books offer additional insights into topics covered in this guide, from historical narratives to technical primers.

## Recommended Websites and Online Resources

- **HowStuffWorks - Electronics** (howstuffworks.com): A beginner-friendly site with articles like "How Semiconductors Work" and "How Microprocessors Work." Great for quick, intuitive explanations and illustrations of basic concepts.
- **All About Circuits - Industry Articles** (allaboutcircuits.com): An educational electronics site that includes tutorials and articles on semiconductor components and manufacturing. Their textbook section has a solid introduction to semiconductor devices and fabrication steps.
- **Semiconductor Industry Association (SIA)** (semiconductors.org): The SIA publishes accessible reports on industry trends, statistics, and policy news. Their "Facts & Figures" and "Insight" sections can give readers an up-to-date picture of the global semiconductor market and technology roadmaps.
- **SemiWiki** (semiwiki.com): A blog and community hub for semiconductor professionals and enthusiasts. SemiWiki features articles on new process technologies, design tools, and industry news written in an accessible way. It's a good place to see commentary on the latest developments (e.g., nodes, chips, mergers) from industry insiders.
- **Intel Newsroom and TSMC Press Releases** (intel.com, tsmc.com): Reading press releases or tech blogs from leading companies can provide insight into cutting-edge tech. They often publish layperson-friendly pieces about new chips or techniques (like Intel's explanation of their new process nodes, or TSMC's announcements on EUV progress).

There are abundant web resources available today. The above list focuses on those with reliable, high-quality information suitable for beginners who want to learn more or stay current. Always exercise caution when browsing forums for information. Stick to known expert communities for accurate info.

## Key Industry Organizations and Initiatives

- **SEMI (Semiconductor Equipment and Materials International)** - An industry association representing equipment and materials suppliers. Semi.org is a global industry association that organizes standards (like the SEMI equipment communication standards) and major conferences (SEMICON) that foster collaboration across the semiconductor supply chain.

- **SIA (Semiconductor Industry Association)** - Industry Association. semiconductors.org. A trade association of semiconductor companies in the U.S., but with global insights. SIA advocates for industry interests in policy and publishes the annual "State of the U.S. Semiconductor Industry" report and technical roadmaps.
- **IEEE Electron Devices Society (EDS)** - Professional Society. ieee-eds.org. A professional organization under IEEE focusing on semiconductor devices, circuits, and related technologies. They organize conferences (like IEDM—International Electron Devices Meeting) and publications where the latest research is shared (though often technical, it drives future industry directions).
- **Global Semiconductor Alliance (GSA)** - Industry Group. gsaglobal.org. A nonprofit organization that brings together a broad range of industry players (fabless, foundry, suppliers) to address common challenges, improve supply chain collaboration, and promote innovation.
- **International Roadmap for Devices and Systems (IRDS)** - Industry Roadmap Initiative. irds.ieee.org. Successor to the International Technology Roadmap for Semiconductors (ITRS). The IRDS is a global roadmap effort sponsored by IEEE, which forecasts technology requirements and potential solutions 15 years out. Their reports provide insight into what might come next (from new materials to novel architectures) and help align academia, industry, and government research efforts.
- **CHIPS for America and EU Chips Act** - Government Initiatives. (commerce.gov and European Commission site). These are major government-funded programs in the USA and the EU, respectively, aiming to boost domestic semiconductor R&D and manufacturing. They signify the strategic importance of chips by providing incentives for building new fabs, advancing research, and workforce development in semiconductors.

These organizations and initiatives are great touchpoints for further exploration. Their websites often have education sections, news updates, and white papers. Simply knowing these bodies exist is also quite helpful. They anchor the semiconductor industry's collaborative and strategic efforts.

# Glossary of Key Terms

**Annealing** Heat treatment process used to repair crystal damage from ion implantation and activate dopants by helping them move into proper positions in the silicon lattice.

**Atomic Layer Deposition (ALD)** A thin-film deposition technique that lays down material one atomic layer at a time, achieving extremely precise thickness control.

**Bandgap** An energy range in a solid where no electron states can exist; in semiconductors, the size of the bandgap determines how easily the material can conduct electricity when energy is applied.

**BGA (Ball Grid Array)** A type of package with solder balls arranged in a grid on the bottom surface, commonly used for high-performance chips like CPUs and GPUs.

**Chemical Vapor Deposition (CVD)** A deposition process where gaseous chemicals react at the wafer surface to form solid films. Widely used for depositing insulators, semiconductors, and some metals.

**Chiplet** A small functional chip designed to be combined with others in a single package to form a larger, more complex processor. Chiplets allow mixing and matching of different pieces (CPU cores, memory, etc.) and often improve yield and flexibility.

**Cleanroom** A specially engineered room with ultra-clean air (very few dust particles). In semiconductor fabs, cleanrooms control temperature, humidity, vibration, and particle count so that delicate chip fabrication processes aren't disrupted by contaminants.

**CMOS (Complementary Metal-Oxide-Semiconductor)** The ubiquitous technology used to build modern integrated circuits, employing pairs of p-type and n-type MOSFET transistors. "Complementary" refers to using both types of transistors in a way that greatly reduces power consumption.

**CMP (Chemical-Mechanical Planarization)** A polishing process that uses chemical slurries and mechanical abrasion to create perfectly flat wafer surfaces between processing steps.

**Czochralski Process** A method for growing large single crystals (such as silicon ingots) by dipping a seed crystal into molten material and slowly pulling it up while rotating. It produces the monocrystalline silicon ingots that are sliced into wafers.


A. D. Watpade, *From Sand to Chip*, Synthesis Lectures on Engineering, Science, and Technology, https://doi.org/10.1007/978-3-032-18439-9

**Damascene Process** A method for creating copper interconnects by first etching trenches in insulator, filling with copper, then polishing away excess copper, leaving copper only in the trenches.

**Die** A single rectangular piece of semiconductor wafer that has been patterned and processed to form an integrated circuit. After wafer fabrication, the wafer is cut (diced) into individual dies (chips).

**Doping** The introduction of specific impurity atoms into an extremely pure semiconductor to change its electrical properties. Doping creates regions of n-type (excess electrons) or p-type (excess holes) silicon, which are crucial for making transistors and other devices.

**EUV (Extreme Ultraviolet) Lithography** Advanced lithography technology using 13.5 nm wavelength light to print the smallest features on cutting-edge chips. Requires special mirrors and vacuum environment.

**Fab (Fabrication Plant)** A factory where semiconductor devices are manufactured. A "fab" contains the cleanrooms and equipment needed for all the processing steps (lithography, etch, etc.) to turn raw silicon wafers into finished chips.

**Fabless** A model of semiconductor company that focuses only on design and outsources the manufacturing. Fabless companies (e.g., Qualcomm, Nvidia) design chips and then have them made in someone else's fab (such companies partner with foundries like TSMC).

**FinFET** Short for "Fin Field-Effect Transistor," a type of 3D transistor where the conducting channel is a thin vertical fin of silicon and the gate wraps around this fin on three sides. This design (first used in production at 22 nm technology) offers better control of the transistor channel compared to older planar designs, reducing leakage and power usage.

**Foundry** A semiconductor company that specializes in manufacturing chips designed by other companies (fabless companies) rather than designing its own products. TSMC is the world's largest foundry.

**FOUP (Front-Opening Unified Pod)** A sealed, portable container used to transport and store silicon wafers within a semiconductor fab. A standard FOUP holds 25 wafers in vertical slots, protecting them from contamination as they move between processing tools, either via automated overhead transport systems or carried by hand.

**Gate-All-Around (GAA) Transistor** An advanced transistor structure where the gate material surrounds the semiconductor channel on all sides. This provides even greater control than a FinFET. Implementations include nanosheet or nanowire transistors. GAA transistors appeared in commercial chips around 2022 (at the 3 nm node).

**Integrated Circuit (IC)** Also simply called a "chip," it's a tiny electronic circuit formed on a small piece of semiconductor material (usually silicon). An IC can contain anywhere from a few to billions of transistors along with other components like resistors and capacitors, all interconnected to perform complex functions.

**Interconnect** The network of metal wires (typically aluminum or copper) that connects transistors together to form functional circuits. Modern chips have multiple layers of interconnect.

**Ion Implantation** A process for doping semiconductors by accelerating ions (charged atoms of dopant elements like boron or phosphorus) in an electric field and shooting them into the silicon wafer. This method precisely controls the dose and depth of dopants and is the primary doping technique in modern chip fabrication.

**Lithography** In chip manufacturing, lithography refers to the process of patterning shapes on the wafer using light (photolithography) or other means. It's analogous to printing: a light-sensitive resist is applied, then exposed through a mask to create the blueprint of circuits that subsequent steps will follow.

**Metrology** The science of measurement in semiconductor manufacturing, using sophisticated tools to measure dimensions, thicknesses, and other critical parameters.

Moore's Law An observation made by Gordon Moore in 1965 that the number of transistors on a chip roughly doubles every 2 years, leading to exponential growth in computing power. While not a physical law, it became a guiding principle for the industry's roadmaps. In recent years, sustaining Moore's Law has become increasingly challenging due to technical and economic limits.

MOSFET (Metal-Oxide-Semiconductor Field-Effect Transistor) The type of transistor used in virtually all modern digital integrated circuits. Has source, drain, and gate terminals with the gate controlling current flow.

Node (Technology Node) A generation of semiconductor manufacturing technology, traditionally named by the smallest feature size (e.g., 7 nm, 5 nm), though modern node names are more marketing terms than actual dimensions.

Photolithography The specific type of lithography that uses ultraviolet light and optical masks to transfer circuit patterns onto a semiconductor wafer. A photoresist (light-sensitive coating) is exposed to light shone through a photomask (which contains the desired pattern). After developing, the remaining resist protects certain areas while other areas can be processed (e.g., etched or doped).

Photomask (Reticle) A quartz or glass plate that carries the circuit patterns to be printed on the wafer. Parts of the mask are opaque (often chrome) and other parts are transparent, forming a stencil. In photolithography, light shines through the mask to project the pattern onto the wafer's photoresist. Modern masks are typically used at a reduced scale (e.g., 4× larger patterns on the mask that get demagnified by the lens).

Photoresist A light-sensitive polymer coating applied to wafers that changes its chemical properties when exposed to light, enabling pattern transfer in photolithography.

Physical Vapor Deposition (PVD) A deposition method that physically removes material from a target and deposits it on the wafer, commonly used for metals. Sputtering is the most common PVD technique.

Planar Transistor The traditional transistor design (used up through the 28 nm node or so) where the transistor's channel lies flat on the silicon surface and the gate covers it from above. It's "planar" because the structure is essentially two-dimensional. Newer 3D designs like FinFETs have replaced planar transistors at advanced technology nodes due to better performance.

Plasma An ionized gas containing charged particles (ions and electrons) used in dry etching and some deposition processes. Often called the "fourth state of matter."

P-N Junction The boundary between p-type and n-type semiconductor regions, fundamental to diodes and transistors. Creates a depletion region that controls current flow.

Quantum Tunneling A quantum physics phenomenon where a particle (like an electron) can pass through a barrier even if it seems it doesn't have enough energy to do so classically. In chips, when transistors become extremely small and their gate oxides very thin, electrons can "tunnel" through insulating barriers, causing leakage current. This is one fundamental limit to making transistors ever smaller.

Selectivity In etching, the ratio of how fast one material etches compared to another. High selectivity means the target material etches much faster than surrounding materials.

Semiconductor A material that is neither a good conductor like metal nor a good insulator like glass, but something in between. Crucially, its conductivity can be modulated. Silicon is the most widely used semiconductor. By doping and applying electrical fields, we can control whether a semiconductor conducts, which is the basis for electronic switches (transistors).

Silicon (Si) The chemical element (atomic number 14) that is the primary material in most semiconductor devices. In its pure crystalline form, silicon forms the wafers that chips are made on. Silicon's semiconductor properties (especially when doped) and its stable oxide (silicon dioxide) make it ideal for integrated circuits. It's abundant in the earth's crust (found in sand as silicon dioxide).

Silicon Dioxide ($SiO_2$) Also called silica or silicon oxide, this glass-like material serves as an excellent insulator in chips and is easily grown on silicon surfaces through thermal oxidation.

Technology Node See "Node" above—same concept, refers to generations of manufacturing technology.

Through-Silicon Via (TSV) A vertical electrical connection that passes through the thickness of a silicon die, allowing signals to travel from one side of the die to the other. TSVs are used in 3D integrated circuits and advanced packaging to stack chips on top of each other and connect them directly, enabling high-density vertical integration.

Transistor The fundamental building block of electronic circuits; essentially an electrically controlled switch. A transistor has at least three terminals: applying a small input voltage/current to one terminal (the gate or base) controls a larger current flow between the other two terminals (source and drain, or emitter and collector). In digital electronics, transistors are used to represent binary states (on/off). There are many types (BJTs, MOSFETs, etc.), but MOSFET transistors are used in virtually all modern digital ICs.

Via A vertical connection that links different layers of metal interconnect in a chip, similar to contacts but connecting metal layers rather than reaching down to transistors.

Wafer A thin, flat disc of semiconductor material (usually silicon) on which microchips are fabricated. Wafers are commonly 150 mm, 200 mm, or 300 mm in diameter (with 300 mm being standard for leading-edge fabs, and 450 mm under research). Dozens to thousands of chips are made on one wafer in a grid pattern, and then the wafer is cut into individual dies after processing.

Wire Bonding A traditional packaging method using fine metal wires (gold, aluminum, or copper) to connect the chip's bond pads to the package leads.

**Yield** In manufacturing, yield refers to the proportion of devices that function correctly out of the total produced. For a semiconductor wafer, yield is often quoted as a percentage of good dies per wafer. Yield can be affected by defects, process variations, and design robustness. High yield is critical for cost-effective production; a process with low yield will make each working chip extremely expensive.

# Index




A. D. Watpade, *From Sand to Chip*, Synthesis Lectures on Engineering, Science, and Technology, https://doi.org/10.1007/978-3-032-18439-9

MIX
Papier aus verantwortungsvollen Quellen
Paper from responsible sources
FSC® C105338

If you have any concerns about our products,
you can contact us on
**ProductSafety@springernature.com**

In case Publisher is established outside the EU,
the EU authorized representative is:
**Springer Nature Customer Service Center GmbH**
**Europaplatz 3, 69115 Heidelberg, Germany**

Printed by Libri Plureos GmbH
in Hamburg, Germany